NEW KEY GEOGRAPHY

Connections

Nelson Thornes
a Wolters Kluwer business

DAVID WAUGH
AND TONY BUSHELL

Key Geography Connections published in 1992 and *New Edition* in 1997 by Stanley Thornes (Publishers) Ltd.
Key Geography New Connections (Third edition) published in 2001 by Nelson Thornes Ltd.

This edition published in 2006 by:
Nelson Thornes Ltd
Delta Place
27 Bath Road
CHELTENHAM
GL53 7TH
United Kingdom

08 09 10 / 10 9 8 7 6 5 4

A catalogue record for this book is available from the British Library

ISBN 978 0 7487 9702 8

Illustrations by Kathy Baxendale, Jane Cope, Nick Hawken, Gordon Lawson, Angela Lumley, Richard Morris, Malcolm Porter, David Russell Illustration, Tim Smith, John Yorke
Edited by Katherine James
Photo research by Penni Bickle and Julia Hanson; research for new edition by Sue Sharp
Original design by Hilary Norman
Page make-up by Viners Wood Associates

Printed in China by Midas Printing International Ltd.

The cover shows photographs of logging in Papua New Guinea (left); a water pump near Alem Kitmama, Ethiopia (middle); and a chemical plant in Lothian, Scotland, UK (right).

The title page shows an offshore wind farm at Middelgrunden, near Copenhagen, Denmark.

Acknowledgements

Cover photos: Digital Vision 15 (NT): left and right; WHO/ P. Virot: middle.

Agripicture Images/ Peter Dean: 27C; Alamy: 27B; Alamy/ Robert Brook: 72A (bottom left); Alamy/ Gary Cook: 117E; Alamy/ Robert Fried: back cover (Brazil); Alamy/ Peter Horree: back cover (Kenya); Alamy/ D Hurst: 96C (top left); Alamy/ Images of Africa Photobank: 107D, 140A; Alamy/ Photofusion Picture Library/ Martin Bond: 135C; Alamy/ Powered by Light/ Alan Spencer: 80A (5); Karl Ammann/ Digital Vision AA (NT): 71C; Andes Press Agency: 123D; Art Directors & Trip Photo Library: 71E, 85 (top and upper middle left), 110A, 114A; Art Directors & Trip Photo Library/ Farmer: 30A; Art Directors & Trip Photo Library/ Greaves: 33C; Art Directors & Trip Photo Library/ Peters: 50C; Baobab Farm Project: 118A, 119B, C & D; Bruce Coleman: 70B (top left, middle and bottom right); Bryan & Cherry Alexander: 128A; Tony Bushell: 6 (bottom left), 7F, 10B, 14B & C, 20A, 24A & B, 85B, 109B; Cambridge Science Park, by kind permission of Trinity College, who established Cambridge Science Park: 58A; Construction Photography: 72A (bottom middle); Corbis/ Derek Croucher: 87B; Corbis/ Gianni Giansanti/ Immaginazione: 46B; Corbis/ Gavriel Jecan: 107C; Corbis/ Lester Lefkowitz: 45B; Corbis/ London Aerial Photo Library: 96B; Corbis/ James Marshall: 4A; Corbis/ Gideon Mendel: 46C; Corbis/ Owaki–Kulla: 5B; Corbis/ Bill Ross: 80A (1); Corbis/ Bobby Yip/ Reuters: 81A (6); Countryside Agency/ Mike Williams: 16A; David Paterson Photography: 6A, 90A; Digital Vision 15 (NT): 65C; Digital Vision PB (NT): back cover (USA); Empics/ AP: 5C, 80A (2); Empics/ PA: 54A, 135C (bottom); Empics/ PA/ John Giles: 18A; Epicscotland.com/ Ross Graham: 89E; Eye Ubiquitous/ J Edkijns: 34; Eye Ubiquitous/ Gibbons: 32A; Eye Ubiquitous/ D Gill: 68B (right); Eye Ubiquitous/ Mowbray: 31C; Eye Ubiquitous/ A Oldfield: 68B (left); Eye Ubiquitous/ Mike Reed: 68B (middle); FLPA: 8 (top left); FLPA/ Christiana Carvalho: 90D; FLPA/ Terry Whittaker: 79C; Geoff Wilkinson/ Geoff Wilkinson Image Library Ltd: 27D; Geosphere Project/ Tom Van Sant: 128C; Getty Images/ H Richard Johnston: 8 (bottom left); L M Glasfibre, Lunderskov, Denmark: 1; Greg Evans Photo Library: 109C; Heritage Image Partnership/ Ann Ronan Picture Library: 44A, 56B; Holt Studios/ Rosemary Meyer: 26A; ICCE Photolibrary/ Malcolm Boulton: 70B (top middle); ICCE Photolibrary/ Mark Boulton: 70B (bottom left); ICCE Photolibrary/ G Faulkener: 37C; ICCE Photolibrary/ Mike Hoggett: 7 (right); Impact Photos/ Roger Scrutton: 46A; James Davis Travel Photography: 109D; Jason Hawkes

Aerial Photo Library: 80A (4); London Aerial Photo Library: 63F; Lonely Planet Images/ Manfred Gottschalk: 89D; Lonely Planet Images/ Chris Stowers: 86A; Magnum Photos/ Peter Marlow: 45C; Ryan McVay/ Photodisc 75 (NT): 96C (middle and bottom left); NI Syndication/ Corbis/ London Aerial Photo Library: 124A; Panos Pictures/ Crispin Hughes: 107B, 114B; Panos Pictures/ Sean Sprague: 90F; Peter Smith Photography: 25C; Photofusion/ Molly Cooper: 103C; Photofusion/ Brenda Prince: 99C; Reportdigital.co.uk/ Philip Wolmuth: 87D; Rex Features: 138A; Rex Features/ Gerard Fritz: 72A (top left); Rex Features/ Alisdair Macdonald: 65C, 87C; Rex Features/ Sipa: 81A (3); Rex Features/ Frank Siteman: back cover (UK); Robert Harding Picture Library/ Robert Estall: 13D; Robert Harding Picture Library/ Michael Jenner: 8 (top right); Rover UK: 54B; Sally & Richard Greenhill: 96C (top, middle and bottom right); Science Photo Library/ Simon Fraser: 72A (top right); Science Photo Library/ Dr Morley Read: 90B; Science Photo Library/ Tom Van Sant/ Geosphere Project, Santa Monica: 92A; Simmons Aerofilms Ltd: 17C, 51D; Simmons Aerofilms Ltd/ Chris Mawson: 15E, 65B; Simon Warner: 90C; Skyscan.co.uk: 78A; Still Pictures/ Mark Edwards: 125C; Still Pictures/ A Ishokon/ UNEP: 125B; Still Pictures/ Jorgen Schytte: 137D, back cover (Bangladesh); Superstock/ Age Fotostock/ P Narayan: 116C; Tony Waltham/ Geophotos.co.uk: 6B,C and bottom right, 25 (inset), 74A; The Travel Library Limited: 106A; Wateraid/ Josh Hobbins: 133E; David Waugh: 75C, 114C, 115D & E, 116A & B, 136C (both), 139D, back cover (China); The Wildlife Trust: 84, 83B (visitor centre).

Map produced from Ordnance Survey mapping with the permission of Ordnance Survey on behalf of HMSO. © Crown copyright (2006). All rights reserved. Ordnance Survey Licence number 100036771: 55C (Landranger 128).

Logos reproduced by kind permission of: Canon, HP Invent, GlaxoSmithKline, IBM, Fujitsu Systems Europe, Apple, Philips and Sony (56A); The National Trust, English Heritage, Woodland Trust, WWF, RSPB, Environment Agency and Greenpeace (68A); WaterAid (133C); Practical Action (139C).

Screenshot (103B): National Statistics website: www.statistics.gov.uk. Crown copyright material is reproduced with the permission of the Controller of HMSO.

Every effort has been made to contact copyright holders and we apologise if any have been overlooked.

Contents

What are rivers and coasts like?

What is this unit about?

This unit explains how river valleys and coasts are affected by erosion, transportation and deposition. It also looks at how coastal erosion can cause severe problems for people.

In this unit you will learn about:

◆ different types of weathering

◆ how material is eroded, transported and deposited

◆ how rivers shape the land

◆ how the sea shapes the coast

◆ the problems of coastal erosion

◆ how coastal erosion may be reduced.

Why is learning about rivers and coasts important?

In Britain we are never very far from a river or the coast. Nearby areas:

◆ are often very attractive

◆ may be used for recreation purposes

◆ provide sites for settlements and industries.

For these reasons, we need to understand river and coastal processes so that we can make the most of their features and manage any problems that might arise. This unit should also help you develop an interest and appreciation of the landscape and scenery around you.

A Angel Falls, Venezuela

B Grand Canyon

C Florida Keys, USA

- For each of the three photos, make a list of words to describe it.

- How do the photos show the power of rivers and the sea?

- Of the places shown in these photos, which:
  - would you most like to visit
  - do you think is the most spectacular
  - do you think looks the most dangerous
  - would you like to know more about?

  Give reasons for your answers.

What is weathering?

There is a great variety of different scenery in the world. Some places are mountainous, some are flat, some can be described as spectacular and others simply as interesting. Geographers call the scenery of a place the **landscape**. Some examples of the world's landscapes are shown in photos **A**, **B** and **C**.

The surface of the earth and the landscapes we see around us not only differ from place to place, but they are changing all the time.

Rain, sun, wind and frost constantly break down the rocks. Great mountain ranges get worn down, valleys are made wider and deeper, and coastlines are changed. The breaking up of the earth's surface in this way is due to **weathering** and **erosion**. Weathering takes place when the rocks are attacked by the weather. Erosion is the wearing away of the land. These two pages show some examples of weathering. Erosion is explained more fully on pages 8 and 9.

A Mount Everest

B Monument Valley, USA

C Guilin, China

Freeze–thaw weathering

This can also be called frost shattering. Water may get into a crack in a rock and freeze. As the water turns to ice it expands and causes the crack to open a little. When it thaws the ice melts and changes back to water. Repeated freezing and thawing weakens the rock and splits it into jagged pieces. This type of weathering is common in mountainous areas where temperatures are often around freezing point.

Water fills a crack in a rock

The water freezes and the crack is made wider

The rock breaks into several pieces

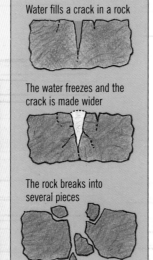

Onion-skin weathering

This happens when a rock is repeatedly heated and cooled. As it is heated, the outer layer of the rock expands slightly and as it cools the rock contracts. Continual expansion and contraction causes small pieces of the rock surface to peel off like the skin of an onion. This type of weathering is common in desert areas where it is very hot during the day but cool at night.

Biological weathering

This is due to the action of plants and animals. Seeds may fall into cracks in the rocks where shelter and moisture help them grow into small plants or trees. As the roots develop they gradually force the cracks to widen and the rock to fall apart. Eventually whole rocks can be broken into small pieces. Burrowing animals such as rabbits, moles and even earthworms can also help break down rock.

Chemical weathering

This is caused by the action of water. Ordinary rainwater contains small amounts of acid. When it comes into contact with rock the acid attacks it and causes the rock to rot and crumble away. The results of this can be seen on buildings and in churchyards where the stone has been worn away or pitted. Water and heat make chemical weathering happen faster, so it is greatest in places that are warm and wet.

Activities

H/W

1 Make a larger copy of diagram **D**.
  a Write in the meaning of weathering.
  b Add labels to the weathering features. **D**

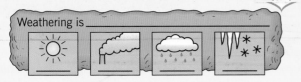
Weathering is _____

2 Copy and complete these sentences.
  a Freeze–thaw weathering is ...
  b Onion-skin weathering is ...
  c Chemical weathering is ...

3 With the help of a labelled diagram, show how freeze–thaw weathering can break up rocks.

4 a Make a larger copy of diagram **E**.
  b Show how root action can break up rocks, by adding the following labels to the correct boxes. Give your diagram a title.

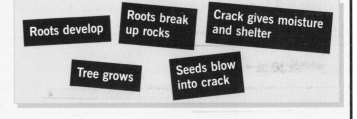
Roots develop | Roots break up rocks | Crack gives moisture and shelter
Tree grows | Seeds blow into crack

E
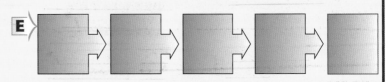

EXTRA

Draw these simple sketches of photos **A**, **B** and **C**. Give each sketch a title and underneath say what type of weathering is likely to be most important there. Give reasons for your answers.

Summary

Weathering is the breakdown of rocks by water, frost and temperature change. Rocks can also be broken down by the effects of plants and animals.

16/9/05

What is erosion ...

Weathering and erosion work together. Weathering breaks up and weakens the surface of rocks while erosion wears away and removes the loosened material. The action of rivers, the sea, ice and wind are the chief types of erosion. Human erosion is also important. Bulldozers and lorries can dig out and move large amounts of soil and loose rock, so changing the landscape. People also remove trees and vegetation which can allow water, wind and ice to erode land more easily.

The work of rivers, the sea, ice and wind are explained in **A** below.

A

Rivers

Every day rivers wear away tiny bits of rock from their bed, and eat into the banks on either side of the channel. This material is carried downstream and deposited when the water slows down. In times of flood large boulders may be loosened and rolled down the river bed.

Ice

A **glacier** is a tongue of ice moving down a valley. Stones and boulders that fall onto it freeze into the ice and act like sandpaper on the rocks beneath. As the glacier moves, it carries the material downwards and at the same time wears away the valley bottom and sides.

Sea

Coastlines are under constant attack by **waves**. During storms each wave hits the rock with a weight of several tonnes. When this is repeated many times, the rock is weakened and pieces break off. **Currents** carry loose material away and deposit it elsewhere.

Wind

Explorers who cross deserts in cars often find their paintwork worn away and their windscreens scratched. This is because the wind picks up tiny particles of sand and blasts them against anything that is in the way. Rocks in desert areas are often eroded into strange shapes by this sandblasting effect.

... and how can it help shape the land?

Look at cartoon **B** on the right. It shows some gardeners who are trying to alter a garden by digging out soil (erosion), moving it in a wheelbarrow (transportation), and dumping it somewhere else (deposition). The more energy they have, the more soil they can dig or transport. When they are tired the digging slows down and they lack the strength to push the barrow, resulting in it toppling over and dumping its load.

On a larger scale, mountains, valleys, plains and coasts are shaped and changed by water, ice and wind. **Erosion** wears away the land, **transportation** moves the material from one place to another, and **deposition** builds up new landforms.

B

Erosion ... ➡ Transportation ➡ Deposition

H/W

Activities

1 a List the following in order of how hard they are. Give the hardest first and the softest last.

steel   chalk   soap   wood

rubber   diamond   plastic

b Put a line under the two you think would be the most difficult to wear down.

c Choose any three from your list and say how they might be worn down.

2 Of the five statements below, three are correct. Write out the correct ones.

- ✓ Weathering is the breakdown of rock by nature.
- ✓ Erosion is the wearing away of rock.
- Weathering and erosion are the same.
- Weathering moves material from one place to another.
- ✓ Erosion includes the removal of loose material.

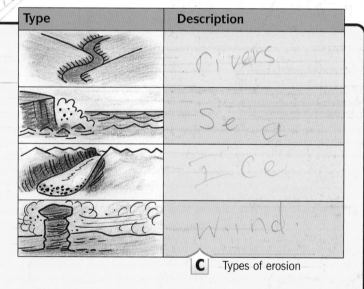

Type	Description
	rivers
	Sea
	Ice
	Wind.

C Types of erosion

3 a Make a large copy of table **C**.

b Add labels to each drawing.

c Write a short description for each type of erosion.

EXTRA

Cartoon **B** shows erosion, transportation and deposition in a garden. How else could this be shown? What about a bulldozer, washing dishes or sandpapering wood? For one of these ideas, or for one of your own, draw a simple labelled cartoon to show how it works.

Summary

Erosion is the wearing away of rock and its removal by streams, ice, waves and wind. Erosion, transportation and deposition help shape the land.

How do rivers shape the land?

Rivers work hard. They hardly stop and they continually erode and move material downstream. They are a major force in shaping and altering the land. Running water by itself actually has little power to wear away rocks. What happens is that the water pushes boulders, stones and rock particles along the river's course. As it does so, the loose material scrapes the river bed and banks and loosens other material. Much of what is worn away is then transported by the river and put down somewhere else. In this way rivers can wear out and deepen valleys. They can also change their shape by depositing material.

The landforms to be seen along a river change as it flows from source to mouth. These two pages explain the features of a river in its upper course which is usually in the hills or mountains. Diagram **A** and photo **B** show how a river cuts out a steep-sided valley that is V-shaped.

A

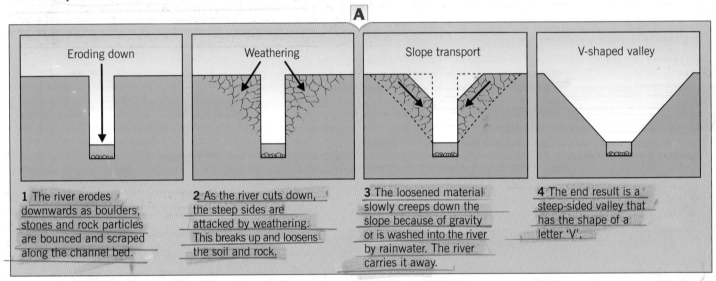

| Eroding down | Weathering | Slope transport | V-shaped valley |

1 The river erodes downwards as boulders, stones and rock particles are bounced and scraped along the channel bed.

2 As the river cuts down, the steep sides are attacked by weathering. This breaks up and loosens the soil and rock.

3 The loosened material slowly creeps down the slope because of gravity or is washed into the river by rainwater. The river carries it away.

4 The end result is a steep-sided valley that has the shape of a letter 'V'.

B

The making of a V-shaped valley

Slopes attacked by weathering

River erodes downwards

Gravity and rainwater move material downwards (slope transport)

River source

V-shaped valley

Rocks and pebbles moved along the bed

Weathered and eroded material transported by river

29/9/08

Table **C** below and sketch **D** give some features of a river and its valley.

C

Source	Where a river starts
Spurs	Ridges of land around which a river winds
Valley sides	The slopes on either side of a river
V-shaped valley	The shape of a valley in its upper course
Channel	The course of a river
River banks	The sides of a river channel
River bed	The bottom of a river channel
Load	Material that is carried or moved by the river

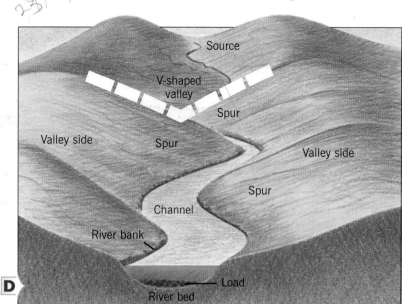

D

Activities

 1 Describe how rivers erode their channels. Include these words in your description:

pushes   scrapes   loosens

moves   drops

2 a Make a large copy of diagram **E**.

  b Show how a valley gets to be V-shaped by describing what happens at ①, ②, ③ and ④.

  c Give your diagram a suitable title.

E

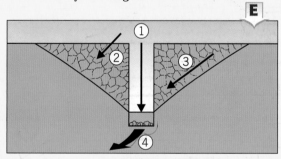

3 a Sketch **F** is a simplified drawing of the river valley shown in photo **B**. Make a copy of the sketch.

  b Add the terms below to your sketch in the correct places. The information at the top of this page will help you.

river channel   river bank   load

valley side   spur   V-shaped valley

F

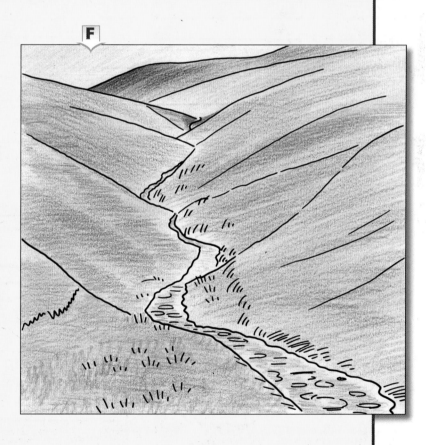

Summary

Rivers erode, transport and deposit material. This helps shape the land. V-shaped valleys are a common feature of a river in its upper course.

What causes waterfalls?

979 m
Angel Falls

244 m
Canary Wharf Tower

Niagara Falls
50 m
High Force
20 m

Waterfalls are an attractive and often spectacular feature of a river. The highest waterfall in the world is the Angel Falls in South America. Its total height is 979 metres. That is about four times the height of Britain's tallest building, the Canary Wharf Tower in London's Docklands. Waterfalls in Britain are much smaller than this (diagram **A**). One of the finest is High Force in the north of England. It has a height of just 20 metres. It is most impressive in times of flood.

Probably the best-known waterfall in the world is Niagara Falls. It lies on the Niagara River which forms part of the border between Canada and the United States.

In this area, a band of hard limestone rock lies on top of softer shales and sandstone. The river flows over the top of the hard rock then plunges down a 50 metre cliff. At the bottom of the cliff the water has worn away the softer rocks to form a pool over 50 metres deep. This is called a **plunge pool**. Down from the falls is the Niagara Gorge. A **gorge** is a valley with almost vertical sides that has been carved out by the river and the waterfall. Photo **D** shows the gorge and waterfall at Niagara.

Sketch **B** shows the Niagara Falls area. The falls here are eating into the cliffs behind the waterfall at nearly one metre a year. The gorge that has been left behind is now 11 kilometres long.

Many waterfalls are formed in the same way as Niagara. They occur when rivers flow over different types of rock. The soft rock wears away faster than the hard rock. In time a step develops over which the river plunges as a waterfall. Water also cuts away rock behind the waterfall. This causes the falls to move back and leave a gorge as it goes. Diagram **C** shows how a waterfall may be worn away by a river.

B

Lake Erie

Niagara River

CANADA

Goat
Island

*Horseshoe
Falls*

**UNITED STATES
OF AMERICA**

*American
Falls*

Original
position
of falls

Niagara Gorge

11 km

CANADA

Hard rock
Soft rock

C

River

Hard rock

Soft rock

Plunge pool

1 Falling water and rock particles or boulders loosen and wear away the softer rock.

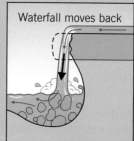

2 The hard rock above is undercut as erosion of the soft rock continues.

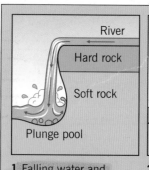

3 The hard rock collapses into the plunge pool to be broken up and washed away by the river. The position of the falls moves back.

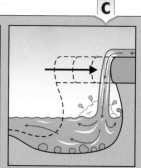

Waterfall moves back

4 Erosion continues and the waterfall slowly eats its way upstream, leaving a gorge behind.

23/9/8

D

Activities

1 Map **E** shows the Niagara Falls area.

a Make an accurate copy of the map.

b Colour the water blue and the land area green.

c Label the following:

USA   Canada   American Falls   Niagara River

Horseshoe Falls   Niagara Gorge   Goat Island

d Draw on and label the original position of the falls.

e The falls have taken 30,000 years to wear back 11 km. Draw on and label where the falls might be 10,000 years from now.

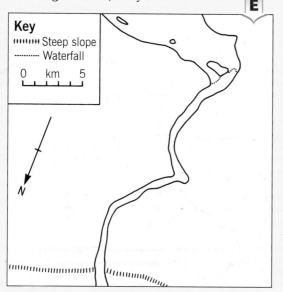

Key
ıııııı Steep slope
‑‑‑‑‑‑‑ Waterfall

0   km   5

N

E

2 a Make a larger copy of diagram **F**.

b Put these labels in the correct places.

Hard rock   Soft rock   Plunge pool

Hard rock breaks off   Eroded material

Undercutting   Waterfall moves back

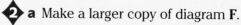

F

3 Sort the phrases in **G** into the correct order and link them with arrows to show how a waterfall may be worn away by a river.

Hard rock collapses   Plunge pool deepened
Soft rock worn away
Waterfall moves back   Hard rock undercut

G

Summary

Many waterfalls are a result of water wearing away soft rock more quickly than the hard rock. As a waterfall erodes back, a gorge may be produced.

13

What happens on a river bend?

Have you noticed that rivers rarely flow in a straight line? Usually they twist and turn as they make their way down to the sea. The only time they are straight seems to be when people interfere with them by building banks or diverting their course.

Bends develop on a river mainly because of the water's eroding power. Think about when you are a passenger in a car and it goes around a corner. You are thrown towards the outside of the curve, often with quite a lot of force. The same happens when a river goes around a bend. The force of the water is greatest towards the outside of the bend. When it hits the bank it causes erosion. This erosion deepens the channel at that point and wears away the bank to make a small **river cliff**. On the inside of the bend, the water movement, or **current,** is slower. Material builds up here due to deposition. This makes the bank gently sloping and the river channel shallow.

Diagram **A** shows what happens on a river bend. At the bottom of the diagram is a **cross-section**. This shows what the river would look like if a slice was cut across it from one side to the other.

A

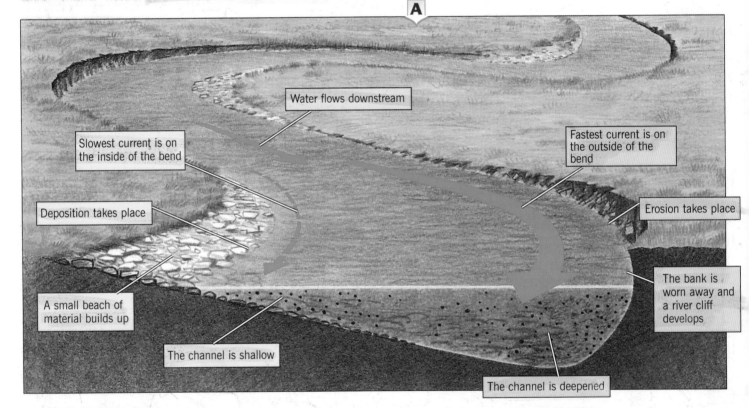

Water flows downstream

Slowest current is on the inside of the bend

Fastest current is on the outside of the bend

Deposition takes place

Erosion takes place

A small beach of material builds up

The bank is worn away and a river cliff develops

The channel is shallow

The channel is deepened

B Deposition on the inside of a river bend

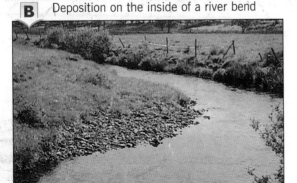

C Erosion on the outside of a river bend

Look at sketch **D** and photo **E**. The river has many bends. These are called **meanders** and are a common feature of most rivers. On either side of the river channel there is an area of flat land called the **flood plain**. This area gets covered in water when the river overflows its banks. Flood plains are made up of **alluvium**, a fine muddy material that is left behind after floods. Alluvium is sometimes called **silt**.

Flood plains are useful to people because they are areas of flat land and have rich fertile soil. This makes them good for building on and for farming.

E

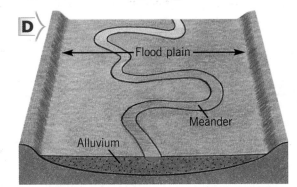

D

Flood plain

Meander

Alluvium

Activities

1 Look at diagram **F**, which is a simple cross-section of a river bend.

a Draw the cross-section.

b Write the labels from **G** in the correct places.

c Give your drawing a title.

d Describe why one side of the river bend is different from the other.

2 a Make simple sketches of photos **B** and **C**.

b For each sketch describe the river feature that it shows.

c Explain how each feature was made.

3 Give the meaning of the terms shown in sketch **D**.

F

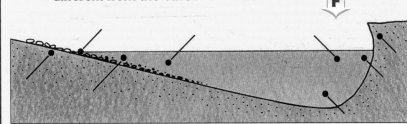

G

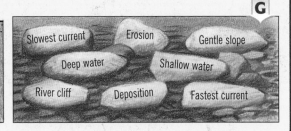

Slowest current   Erosion   Gentle slope

Deep water   Shallow water

River cliff   Deposition   Fastest current

EXTRA

1 Write down two reasons why the flood plain of a valley is good for farming.

2 Give one problem of farming the flood plain. Suggest what could be done to reduce the problem.

Summary

A river course is seldom straight. It usually has many bends which cause it to meander down its valley. The outside of a river bend is worn away by erosion while the inside is built up by deposition.

How does the sea shape the coast?

The sea is never still. On quiet days the movement is slow and gentle, and the sea is flat and almost calm. On stormy days large waves crash against the shore. These large waves have such force that they can drive a ship against the rocks or smash up sea defences and piers. The sea can also wear away the coast and move bits of rock and sand from one place to another. This ability to erode, transport and deposit material produces many interesting coastal landforms.

Erosion landforms are made by the wearing away of the coast (photo **A**). In stormy conditions the sea picks up loose rocks and throws them at the shore. This bombardment undercuts cliffs, opens up cracks and breaks up loose rocks into smaller and smaller pieces. Areas which have soft rocks are worn away more easily than those with hard rocks. The soft rock areas become **bays** and the hard rock areas become **headlands**. A bay is an opening in the coastline. A headland is a stretch of land jutting out into the sea.

A Old Harry Rocks, the Foreland, Dorset

Sketch **B** shows how a headland is eroded by the sea and how other landforms develop.

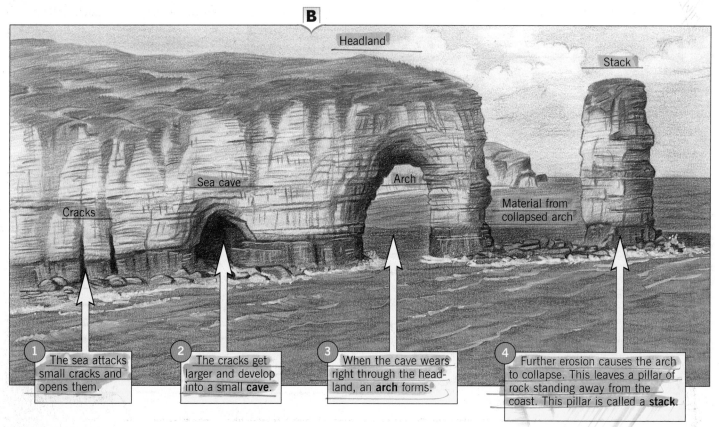

B

Headland

Stack

Sea cave

Arch

Material from collapsed arch

Cracks

1 The sea attacks small cracks and opens them.

2 The cracks get larger and develop into a small **cave**.

3 When the cave wears right through the headland, an **arch** forms.

4 Further erosion causes the arch to collapse. This leaves a pillar of rock standing away from the coast. This pillar is called a **stack**.

Beaches are one of the most common features of our shoreline. They are formed when material worn away from one part of the coast is carried along and dropped somewhere else. A beach is an example of a **deposition landform**.

A **spit** is a special type of beach extending out into the sea. It is a long finger of sand and shingle that often grows out across a bay or the mouth of a river.

Photo **C** and drawing **D** show Spurn Head spit. The spit is 6 kilometres long and forms a sweeping curve that stretches halfway across the mouth of the River Humber. It is continually changing its shape as new material is deposited and old material is worn away or moved elsewhere.

1. Erosion of coastline north of Spurn Head
2. Eroded material transported down the coast by sea currents
3. Material dropped where coastline changes direction
4. Spit grows out from coast as more material builds up
5. End of spit curved by action of the waves
6. Marshland formed behind spit

C The making of Spurn Head spit

Activities

1 a Make a sketch of photo **A**.

b Label these features on your sketch:

crack   arch   cave   stack

material from a collapsed arch

c Explain how the arch was formed.

d Draw a dotted line to where there was once another arch.

H/W

2 a Copy the drawings of Spurn Head spit shown below (**D**).

b With the help of the drawings, describe the formation of the spit.
You could add information to the drawings as well as writing an explanation underneath. Include these terms in your description.

erosion   Flamborough Head   Spurn Head   6 km

currents   deposition   transportation   grows

D

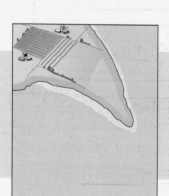
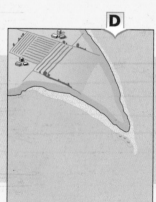
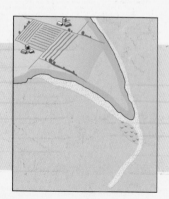

Summary

The coastline is always changing its shape. Some parts are being worn away by erosion while other parts are being built up by deposition.

What is the coastal erosion problem?

Coastal erosion can cause severe problems. Agricultural land may be lost, buildings destroyed and transport links put in danger. The east coast of England has some of the fastest-eroding coastlines in Europe. One of the areas in most danger is Holderness in Yorkshire where the sea is eroding the land at about 2 metres every year.

Over the last 2,000 years, several villages and farms have disappeared into the North Sea as the Holderness coastline has gradually moved back. The village of Mappleton is the latest victim. Already several houses have fallen into the sea, some farm buildings have become unsafe and the coastal road is all but lost (photo **A**).

Many people in the Holderness area are worried about their future. They are afraid of losing their homes and, in some cases, their livelihoods. Several farms are threatened and seaside resorts like Hornsea and Withernsea, where many people work, are also in danger. A cliff-top gas plant for gas piped from the North Sea is also at risk.

A

Holderness

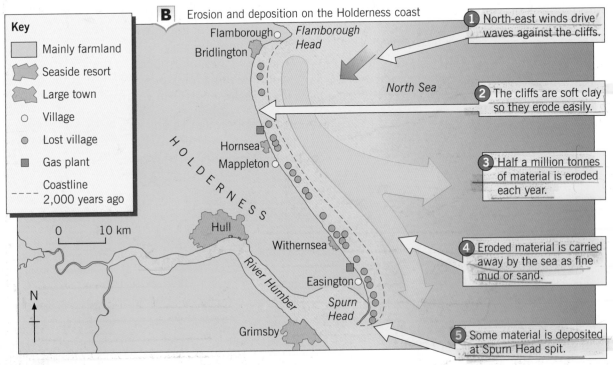

B Erosion and deposition on the Holderness coast

Key

	Mainly farmland
	Seaside resort
	Large town
○	Village
●	Lost village
■	Gas plant
---	Coastline 2,000 years ago

0   10 km

N

Flamborough
Bridlington
Flamborough Head

North Sea

HOLDERNESS

Hornsea
Mappleton

Hull

Withernsea

River Humber

Easington

Spurn Head

Grimsby

1 North-east winds drive waves against the cliffs.

2 The cliffs are soft clay so they erode easily.

3 Half a million tonnes of material is eroded each year.

4 Eroded material is carried away by the sea as fine mud or sand.

5 Some material is deposited at Spurn Head spit.

All around Britain's coast, the cliffs are being worn away by waves and weathering. It happens all the time. The process never stops. But why is erosion such a problem along the Holderness coast and why is it worse here than on any other part of the coastline?

The main reason is that the rock in this area is soft boulder clay left by the glaciers during the last Ice Age some 10,000 years ago. It is easily worn away by weathering and the constant pounding of waves. Drawing **C** shows the processes at work along the Holderness coast.

Rain soaks into cracks, dissolves minerals and weakens the structure.

In winter, water in the cracks freezes, the ice expands and opens up the cracks.

Rain makes clay very slippery and heavy. Huge sections just slide down the slope.

Waves crash into the cliff and gradually wear away the rock.

Pebbles and rocks thrown with great force by the waves undercut the cliff.

The cliff collapses and eroded material is washed away by the sea leaving no protection.

C Cliff collapse at Holderness

Activities

1 Look at map **B**.

 a How far has the coastline moved back in 2,000 years?

 b How many villages have been lost?

 c Suggest why the coast near Grimsby has not been affected by erosion.

2 Look at maps **D** and **E**.

 a How far has the coastline at Mappleton moved back over 30 years?

 b How many buildings have been lost?

 c Which buildings should soon be closed down?

3 Make a list of the problems caused by coastal erosion in the Holderness area. Sort the problems under the headings:

 Buildings   Industries   Employment

4 Many people like Mappleton as a place to live. Suggest three reasons **for** living there and three **against**.

H/W

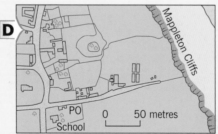

D Mappleton 30 years ago

PO
School
0   50 metres

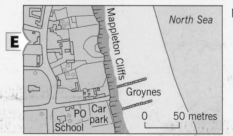

E Mappleton today

North Sea

Mappleton Cliffs

Groynes

PO   Car park
School
0   50 metres

Summary

Erosion is a problem for many of our coastal areas. It causes land loss and may destroy property, transport links and industries. It can also result in job losses.

How can coastal erosion be reduced?

Protecting coasts can be both difficult and expensive. Where valuable land or property is under threat from the sea, local authorities try to slow down or prevent erosion. Since the year 2000, over £1 billion has been spent defending Britain's coastline. Photo **A** and drawing **B** show some of the methods that can be used.

Unfortunately, putting in new sea defences is not always the best solution. Geographers know that protecting one part of the coast can cause even worse problems further along the same coastline.

At Mappleton, for example, new sea defences built in 1992 have helped protect the village but have led to greater erosion of the cliffs to the south.

Many people nowadays are actually against building coastal defences. They support schemes which work with nature rather than against it. Schemes like this, they say, do less damage and help retain wildlife and the quality of the natural environment.

A Sea defences

B Some methods used to help reduce coastal erosion

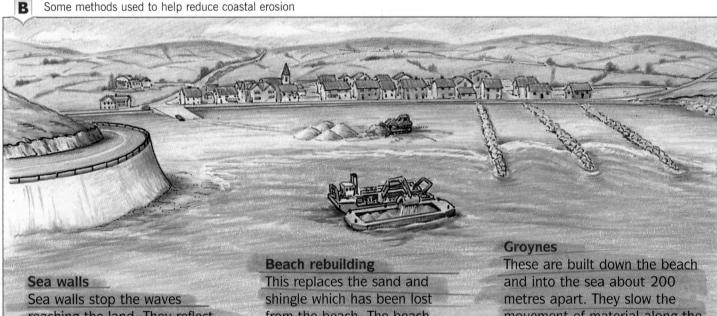

Sea walls
Sea walls stop the waves reaching the land. They reflect the waves back to sea but this can wash away the beach. They give good protection but are expensive and may need to be repaired in time.
Cost: *about £7 million per km*

Beach rebuilding
This replaces the sand and shingle which has been lost from the beach. The beach absorbs wave energy and is a good defence against the sea. It protects the land or sea wall behind the beach and looks more natural.
Cost: *about £2 million per km*

Groynes
These are built down the beach and into the sea about 200 metres apart. They slow the movement of material along the coast and help build up the beach. The beach then helps protect the land. Rock groynes are expensive.
Cost: *about £1.5 million per groyne*

The Holderness coast

So what should be done with the Holderness coast? There is no easy solution but what experts do agree on is that any plan must look at managing the whole coastline rather than just parts of it. The following are three possible solutions.

1 Build sea defences along the whole 60 km of coastline. This would be expensive but would be the choice of most local residents who argue that everyone along the coast deserves to be protected. One problem could be that it would cut off the supply of sand to Spurn Head, which might disappear altogether.

2 Protect the main towns and allow the sea to erode the land between these places into small bays, as shown on map **C**. Beaches would form in the bays which would help protect the coast from further erosion. Some material would continue to be transported down to Spurn Head and beyond.

3 A third solution might be to do nothing at all and let nature take its course.

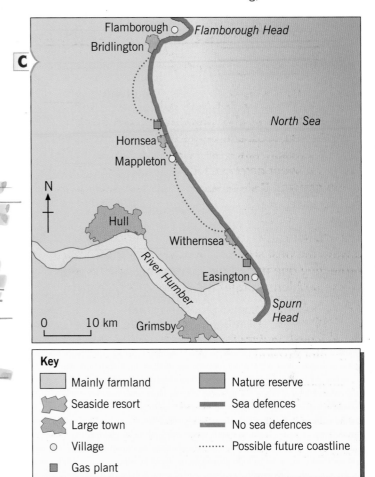

C

Key

☐ Mainly farmland	☐ Nature reserve
Seaside resort	▬ Sea defences
Large town	▬ No sea defences
○ Village	⋯⋯ Possible future coastline
☐ Gas plant	

Rip-rap

This is a mixture of large boulders and concrete blocks which protect the coast by breaking up the waves. They don't protect cliffs as well as a sea wall but they do help retain the beach. They can look ugly and make beach access difficult.

Cost: *about £3 million per km*

Activities

1 Look at the coastal defences in drawing **B**.

   a Which do you think would be:
   - easiest to build
   - most attractive to look at
   - cause fewest problems for people using the beach?

   b Work out the cost of protecting all 60 km of the Holderness coast for each of the methods shown.

2 What are the arguments against protecting the whole coastline?

3 **a** Which of the three solutions would you choose for Holderness?

   b Describe the scheme.

   c Give the advantages and disadvantages of the scheme.

Summary

Protecting coasts is not easy. There are arguments for and against trying to protect the coastline from erosion.

The river enquiry

The Pennine Way is one of Britain's best-known long distance walks. It stretches some 410 km (256 miles) from the Derbyshire Peak District northwards to the Cheviots and into Scotland. One of the walk's most interesting sections is in Upper Teesdale. Here the trail follows the River Tees through some wild and impressive scenery. At one point it overlooks High Force, one of Britain's finest waterfalls.

Every year the Upper Teesdale area receives more and more visitors. Some people have suggested that a small pamphlet explaining the features of the area would be helpful. This would enable walkers and tourists to understand their surroundings and help them enjoy their stay in the area. Your task is to produce the pamphlet by answering the enquiry question below.

To do this you will need to describe the main features of the area and find out how they were formed. You could work by yourself, with a partner or even in a small group. You might be able to use a computer to word process your work and make it look more professional. Pages 8 to 15 of this book will be helpful to you.

How does the River Tees shape the land?

Your aims for this piece of work are:
◆ to answer the enquiry question
◆ to produce an information pamphlet for use by walkers and other visitors to Upper Teesdale
◆ to describe and explain the main physical features of the area
◆ to make your pamphlet attractive, interesting, helpful and simple to understand.

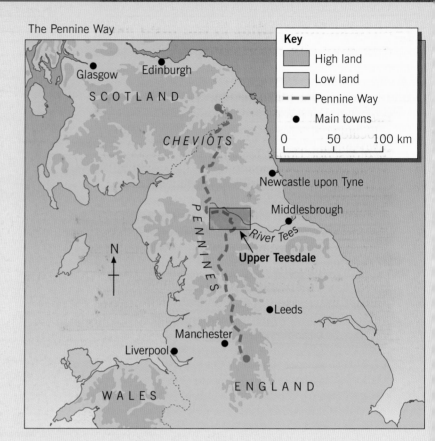

The Pennine Way

Key
High land
Low land
--- Pennine Way
● Main towns
0   50   100 km

1 Introduction – what is the area like, and how can rivers shape the land?

You will need to use maps and writing here. Diagrams and even cartoons might also help.

a First say what the Pennine Way is and show where it is located.

b Next describe Upper Teesdale. You will need to:
◆ say whereabouts it is
◆ describe the scenery
◆ suggest why it is a popular area for visitors.

The text and map on page 22 and the sketch below will help you with this. Try to use the words on the notice board in your description.

c Finally, describe how rivers can shape the land through erosion, transportation and deposition processes.

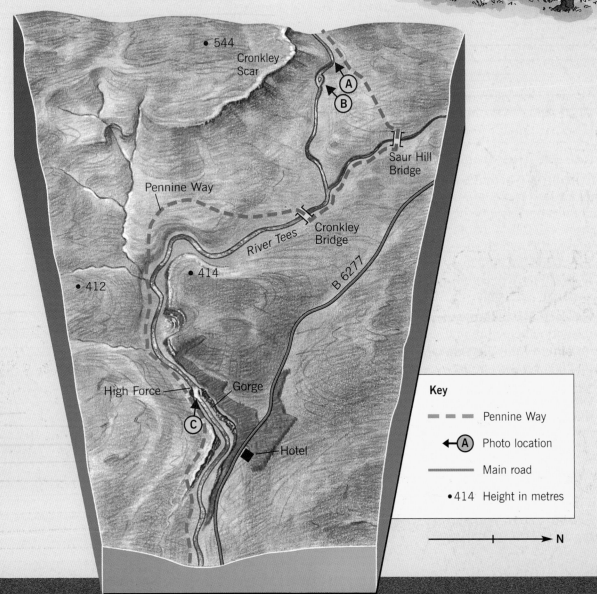

Key

- - - Pennine Way

◀─Ⓐ Photo location

───── Main road

•414 Height in metres

A

The river on this bend is about 25 metres wide with a depth of no more than half a metre. The deepest water is on the right of the photo.

2 What are the main landform features and how were they formed?

a Look carefully at photos **A**, **B** and **C**, which show a river bend, an area that floods, and High Force waterfall and gorge.

b For each one in turn, describe the feature and explain the river processes that helped form it. You could use labelled sketches, cross-section drawings, block diagrams and writing here.

Try to present your work in a clear and interesting way. It is important that your readers find your work attractive and are able to understand fully what you are showing them.

B

At times of flood, the river here is almost 300 metres wide. The valley floor is made up of alluvium (silt) and boulders of all sizes.

At High Force the River Tees plunges down a vertical cliff of over 20 metres. The wearing away of the rock by the falls has left a steep-sided gorge. This is some 700 metres in length.

C

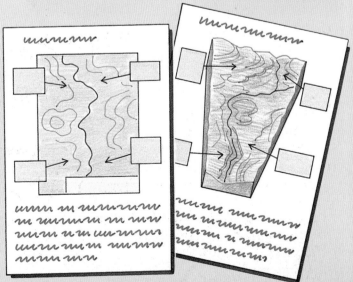

Hard whinstone

Soft limestone

Plunge pool

3 Conclusions

This should be a summary to go at the end of your information pamphlet. There could be two parts to it. Together they should answer the question set by the enquiry question.

a First you need to describe briefly the three main landform features of the area, and show their location. You could do this by adding labels to a map or sketch like the one on page 23. You should write no more than about 20 words for each label.

b Secondly you should write a paragraph to explain how the River Tees has helped shape the land in this area.

What is farming like in Britain?

What is this unit about?

This unit is about farming in Britain. It looks at the type and distribution of farming, how farming affects the landscape and how the farming industry has changed in recent years.

In this unit you will learn about:

◆ the main types of farming in Britain

◆ arable and hill sheep farms

◆ the distribution pattern of farming in Britain

◆ how farming has changed the landscape

◆ changes in farming and their effects.

Why is learning about farming important?

Although most of us live in towns and cities, much of Britain is still countryside. In these areas:

◆ farmers manage more than two-thirds of the land

◆ farmers provide most of the country's food

◆ farming is usually the main source of jobs.

For these reasons we need to have a knowledge and understanding of farming, because:

◆ it helps provide jobs and generates wealth

◆ it affects what we eat and how much we spend on food

◆ it alters the appearance of our countryside

◆ it can have a major effect on wildlife.

A Gower, Wales

B | Cumbria

- In what ways has farming changed the look of the countryside in photo **A**?
- What problems does the farmer in photo **B** face?
- What is happening in photo **C**? What do you think is good and bad about this?
- What is good about the market in photo **D** both for farmers and shoppers?

C | Cornwall

D | Bristol

SUNNYFIELDS ORGANIC FARM

What types of farming are there in Britain?

Farming, or **agriculture**, is the way that people produce food by growing crops and raising animals. There are three main types of farming in Britain.

◆ **Arable** is the ploughing of land and the growing of crops.

◆ **Pastoral** is leaving the land under grass for the grazing of animals.

◆ **Mixed farming** is when crops are grown and animals are reared in the same area.

Farming, especially in Britain, is big business. Farmers must carefully choose the best type of farming for the place where they farm. Deciding which is the best type depends upon several physical and human factors.

◆ **Physical factors** are climate, relief and soils.

◆ **Human factors** include farm size, technology, machinery, distance from markets, and transport.

A **Arable farming**

I grow crops like wheat, barley, potatoes and carrots. They need flat land with deep, fertile soil and a warm climate. There must be some rain, but not too much. Sunshine helps the crops to ripen. Modern machinery helps us with our work. We use it to plough the land and harvest crops.

Arable farmers who grow crops that are heavy to transport (such as potatoes) or soon go bad (such as strawberries) must be near to the market.

B **Hill sheep farming**

Sheep often live in areas where soils are thin and have little goodness in them. They feed mainly on grass, heather and other poor-quality vegetation. Most hill sheep farms like mine are very large but make little use of modern machinery.

On our farm we make a living by selling wool, lamb and mutton at the local market. We live in the hills but sheep are very tough and can graze land that is too steep to raise cattle or grow crops. They can survive in any climate in Britain.

C Cattle farming

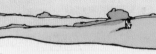

We have two herds of cows. One is for milk and dairy products and the other one for beef. Cows need land that is quite well drained and is either flat or gently sloping. They feed on grass, which grows best in a moist, warm climate.

Dairy farms need to be next to good roads so that the milk can be quickly transported to nearby markets. Refrigerated milk lorries help to keep milk fresh and enable it to be transported longer distances.

D Mixed farming

We grow crops and raise livestock. Our farm is ideal for mixed farming as most of it is flat or gently sloping. This is good for the cattle and makes it easy to use modern machinery when ploughing the fields or harvesting crops.

Mixed farms need good-quality soil and a climate that is neither too wet nor too dry. These factors help grass and crop growth. Good roads and a nearby market where we can sell our produce, are also important.

Activities

1 Match the following beginnings to the correct endings.

Arable farming — is rearing animals

Pastoral farming — is rearing animals and growing crops

Mixed farming — is growing crops

2 Diagram **E** shows some newspaper headlines about problems which affect farming. Copy table **F** below and sort the headlines in **E** into the correct columns.

E >

Wet summers cause disease in potatoes

Ferry strike – fruit rots in French ports

Sheep are stranded in snowdrifts

Farms are too small for family to earn a living

Rise in petrol prices

Crops ruined by flooding

Fruit blossoms killed by frost

High electricity prices – dairy farmers with milk machines hit hard

Daffodils flattened by strong winds

Problems which affect farming	
Physical	Human

F

Summary

The three main types of farming in Britain are arable, pastoral and mixed. The type of farming that is best for an area depends on several physical and human factors.

What is an arable farm like?

Photo **A** and map **B** show Hawthorn Farm, a typical arable farm in East Anglia. The land here is very flat and slopes gently eastwards towards a river. Most of the fields have deep and fertile soils. These are easy to work and are well drained, although occasional flooding restricts the use of land near to the river.

East Anglia is one of the driest parts of Britain. Much of the rain falls in the growing season when it is most needed. Summers are generally warm with plenty of sunshine to ripen the crops. The cold winters have hard frosts which help to kill diseases and break up the soil, which assists ploughing.

The main problems for farmers in the area are occasional high winds and summer thunderstorms that can ruin the crops just before harvest time. Farmers cope with unusually dry summers by **irrigating** the land through a system of ditches, pipes and pumping units. Pumps are also used to move water from the fields and into the river when flooding occurs. Flooding is a problem in the area because the land is so flat and low-lying. Many farmers are concerned that floods happen more often now than they did in the past.

A

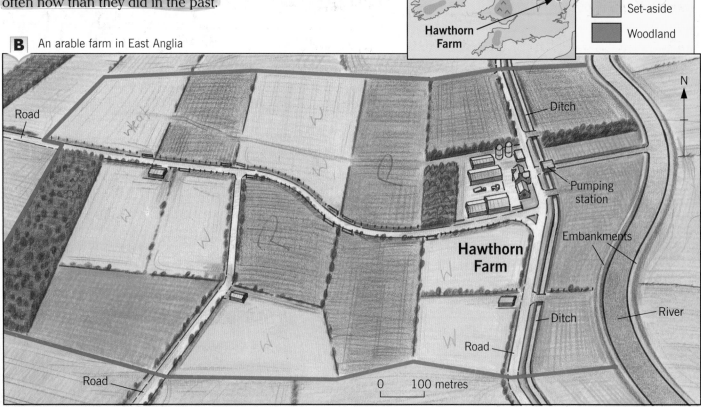

Key

Wheat
Potatoes
Sugar beet
Peas
Set-aside
Woodland

Hawthorn Farm

B An arable farm in East Anglia

Road

Ditch

Pumping station

Embankments

Hawthorn Farm

Ditch

River

Road

Road

0   100 metres

Hawthorn Farm is large and efficient, and uses modern methods. The farm is owned by a company and run by a professional manager. The company has the money needed to invest in new farming methods. Farming like this is called **agribusiness**.

Much use is made of machines and chemicals. Hawthorn Farm has five tractors, two combine harvesters, muck spreaders, sprayers, ploughs, seed drills and a grain drier. There are six full-time labourers, and several casual workers are employed at different times of the year.

The farm makes its money, of course, from selling crops. Some crops, such as peas, go direct to Bird's Eye, a large food-processing company. Others are kept in huge refrigerated stores and sold to wholesalers and supermarkets such as Tesco and Safeway when prices are favourable.

The Government and European Union provide annual cash payments and **subsidies** for arable farmers. These come in the form of grants and loans which help the farmer invest in modern methods and to grow more food. Without subsidies, large numbers of farmers would go out of business.

C

Activities

1 Look at diagram **D**, showing the farmer's year.

  a Which are the three warmest months?

  b During which months is the land being prepared for growing wheat?

  c During which months is the wheat growing?

  d When would be a good time for arable farmers to go on holiday? Suggest reasons for this.

2 Make a fact file to show the main features of an arable farm. Use the headings shown here.

Arable farm
- Location ...
- Relief ...
- Soils ...
- Climate ...
- Method ...
- Use of machinery ...
- Source of income ...
- Difficulties ...

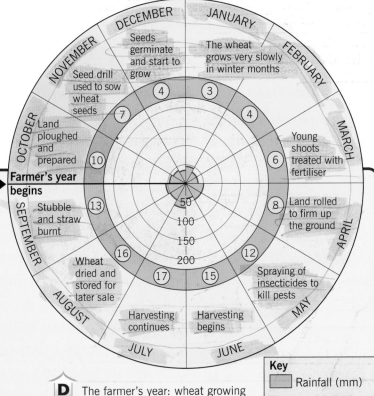

D The farmer's year: wheat growing

Key
Rainfall (mm)
Total rainfall 545 mm
(15) Temperature (°C)

Summary

The climate, relief and soils of East Anglia are ideal for growing crops. Modern farming methods and subsidies help make arable farming more efficient and successful.

What is a hill sheep farm like?

Beckside Farm in the English Lake District is a typical hill sheep farm. Most of the farm is made up of high fell and steep and rocky hillside. The soils here are poor and will only support rough grasses and heather. The farm has only a small area of flatter, low-lying land. This has deeper soils but is often wet and difficult to cultivate.

Modern machinery is little used at Beckside Farm. The valley sides are too steep and inaccessible for tractors and equipment, whilst the valley floor, with its small fields and often waterlogged land, restricts the use of heavy machinery.

The climate in the area can be difficult for farming. Rainfall is heavy and low cloud and mist are common. Summers are cool and winters can be cold and windy. Snow may lie on the higher ground for several weeks. Even sunshine amounts are lower than in the rest of the country.

A

B A hill sheep farm in the Lake District

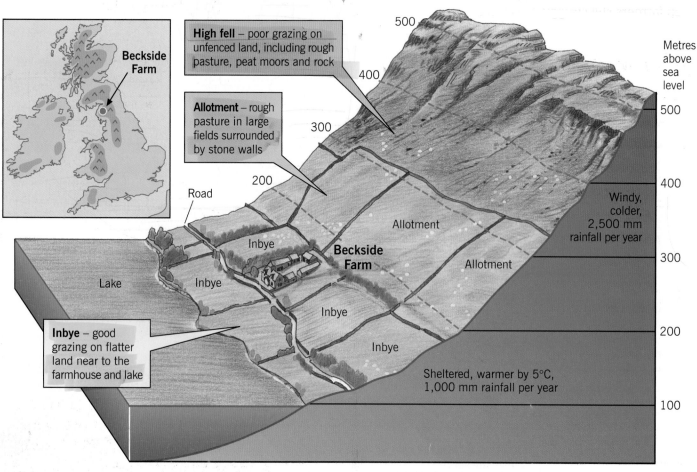

Beckside Farm

High fell – poor grazing on unfenced land, including rough pasture, peat moors and rock

Allotment – rough pasture in large fields surrounded by stone walls

Road

Inbye

Inbye

Lake

Inbye

Inbye – good grazing on flatter land near to the farmhouse and lake

Allotment

Beckside Farm

Allotment

Inbye

Inbye

Metres above sea level

500

400

300

200

100

Windy, colder, 2,500 mm rainfall per year

Sheltered, warmer by 5°C, 1,000 mm rainfall per year

The sheep on Beckside Farm spend most of their time on the high fells. Life here is hard and only strong, healthy sheep survive. In winter, when conditions on the mountainside are uncertain, weaker sheep are brought down to lower pastures.

The small area of lowland is well used. The fields are ploughed each year and fertiliser added to improve the soil's fertility and structure. Oats, barley and turnips are grown during the summer and stored in the large barns. These are then given to the sheep as winter feed.

Beckside Farm earns most of its money in the autumn when lambs and four-year-old sheep are taken to the nearby market at Penrith. Wool is sold in summer but is no longer a main source of income.

Making a living on a hill sheep farm is difficult, so both the Government and the European Union provide annual cash payments and **subsidies** for the farmer. These come mainly in the form of grants and loans. Even with this help, many hill sheep farmers still struggle to survive.

C

In the harsh, rough grazing of the fells, only hardy breeds of sheep can be reared. The **Herdwick** is the toughest breed of sheep in Britain and can cope well with wet and cold conditions. It has a thick grey fleece and is long-lived. The **Swaledale** is now the most common breed in the Lake District. It is hardy and has good lambing and breeding ability.

Activities

1. Look at diagram **D**, showing the farmer's year.

    a Which are the four warmest months?

    b Which are the four wettest months?

    c How much rainfall is there in January?

    d When do the sheep need extra food? Suggest reasons for this.

2. Make a fact file to show the main features of a hill sheep farm. Use the headings shown here.

Hill sheep farm
- Location ...
- Relief ...
- Soils ...
- Climate ...
- Method ...
- Use of machinery ...
- Source of income ...
- Difficulties ...

DECEMBER — Winter feed started — 3
JANUARY — Winter feed given — 3
NOVEMBER — Dipping and mating — 6
FEBRUARY — Winter feed given — 3
OCTOBER — Surplus ewes and lambs sold — 8
MARCH — Flock gathered — 5 — Vaccinations
SEPTEMBER — New lambs taken from sheep — 12
APRIL — Lambing begins — 7
— 50 — 100 — 150 — 200
AUGUST — Fleeces sold — 14
— Sheep returned to fell — 9
— Shearing and dipping — 14 — 12
MAY
JULY — JUNE

Key
Rainfall (mm)
Total rainfall 1,880 mm
⑭ Temperature (°C)

D The farmer's year: hill sheep farming

Summary

Harsh conditions and poor land quality make farming difficult in upland areas. Hill sheep farming is common in places where it is not possible to raise cattle or grow crops.

What is the pattern of farming in Britain?

We have seen how the type and methods of farming are influenced by physical and human factors. As these factors can change within a few kilometres, so too can the type of farming. To draw a detailed and accurate map to show where the main farming types are found in Britain would be very complicated. It is often more meaningful to draw a simplified, or generalised, map instead.

Map **A** is generalised. It is easy to draw and to understand. However, if you look at an atlas map showing 'farming in Britain' you will see several important differences which the generalised map cannot show. These include arable farming in eastern Scotland, sheep on the Pennines and market gardening in Lancashire.

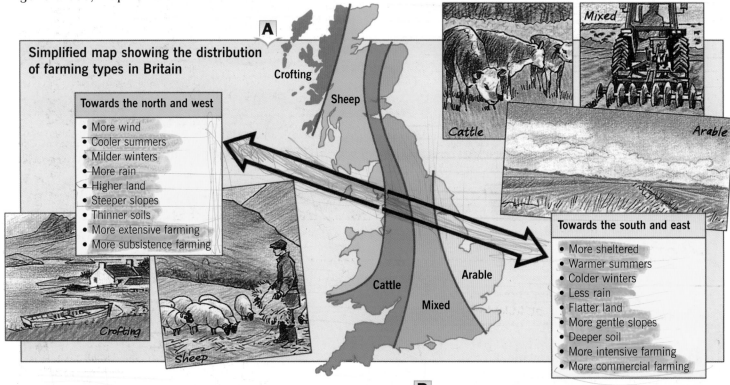

A

Simplified map showing the distribution of farming types in Britain

Crofting

Sheep

Towards the north and west

- More wind
- Cooler summers
- Milder winters
- More rain
- Higher land
- Steeper slopes
- Thinner soils
- More extensive farming
- More subsistence farming

Crofting

Sheep

Cattle

Arable

Mixed

Mixed

Cattle

Arable

Towards the south and east

- More sheltered
- Warmer summers
- Colder winters
- Less rain
- Flatter land
- More gentle slopes
- Deeper soil
- More intensive farming
- More commercial farming

How is this distribution pattern changing?

The most recent changes have resulted from decisions made by the European Union (the EU) in Brussels. For example, since the mid-1970s there has been a rapid increase in farm cultivation of **oilseed rape** in Britain (photo **B**). This crop is grown for several reasons:

◆ Britain and the EU are short of oilseed.

◆ Oilseed is used to make margarine and cooking oil.

◆ By producing our own oil, shop prices are kept lower.

◆ Farmers get a high price for oilseed.

◆ After harvesting, the remains can be fed to cattle.

◆ Rape puts goodness (nitrogen) back into the soil and so is an ideal rotation crop.

B

Farming has also changed in other ways. In the past farmers received money, or **subsidies**, for certain types of farming regardless of whether it was best for them or their land. Since 2005 this has changed and farmers now receive a single payment each year based on the size of the farm. This has given farmers greater freedom to choose exactly what they want to do with their land.

Additional money is also available to farmers who take on environmentally friendly schemes such as those shown in drawing **C**.

C

Leave a protection zone along hedges so that wildlife can thrive.

Avoid cutting hedges during the bird-nesting season.

Look after stiles and gates on public footpaths.

Public Footpath

Plant trees on their land as part of the **Farm Woodland Scheme**.

Set-aside land for non-farming use such as natural vegetation.

Activities

1 Five types of farming are shown on map **A**. Match each type with one of the following descriptions.
- Animals kept for milk and meat
- A subsistence form of farming
- Growing crops
- Growing crops and rearing animals
- Hill farming

2 Match the numbers on map **D** with the following farming types. Some may be used more than once. Map **A** will help you.

Sheep	Cattle	Mixed
Arable	Crofting	

3 Imagine that you are an arable farmer and you have been told that you are no longer going to receive subsidies. Would you:
- still keep on growing crops
- 'set-aside' some land to be left as grass, or
- plant several hectares of trees?

Give reasons for your answers.

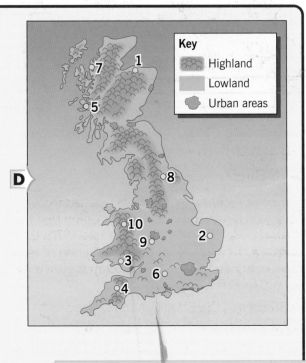

D

Key
- Highland
- Lowland
- Urban areas

Summary

It is possible to recognise a generalised pattern of farming in Britain. This pattern has taken centuries to develop and is still changing.

How has farming changed the landscape?

Most of southern Britain was once covered in forest and marshland. As farming developed the forests were cleared and the marshes were drained. Later, hedges were planted and stone walls built. These created the fields which we now think of as being typical of farming areas.

Recently many more changes have taken place in our landscape. Sketch **A** shows some of the changes which have taken place in the last 60 years.

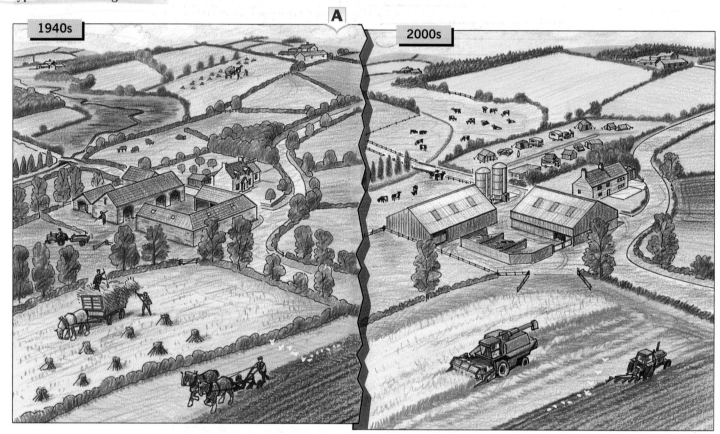

A

1940s

2000s

It is the farmers' job to try to produce more of our food. To do so they have drained **wetlands** and cleared **hedgerows**. These two changes have upset many people.

Wetlands are marshy areas. They also form one of the last of our natural environments, providing homes for a wide variety of wildlife. As wetlands are drained, mainly for pastoral farming, the habitats of many birds, insects, animals and plants are destroyed.

Hedges were planted to stop animals from wandering and to show boundaries of land. Many people consider hedges to be very important to the environment. While farmers may share these views, it is their land which is taken up by hedges and their job to look after them. Since 1945, about one kilometre of hedge out of every four has been cleared (diagram **B**). Every cleared kilometre gives one extra hectare of land. (Remember: some farms in India are only that size!)

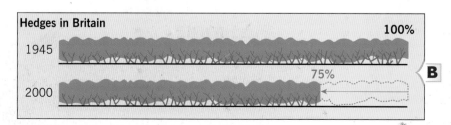

Hedges in Britain

1945   100%

2000   75%   **B**

Well looked after hedges are attractive

Hedge roots hold the soil together and reduce erosion

Hedges provide a home for wildlife

Cutting hedges costs the farmer time and money. A hedgecutter costs over £7,000

Hedges take up space which could be used for farmland

Hedges get in the way of big machinery in fields

C Advantages and disadvantages of hedges

D

Photo **D** shows **soil erosion** in central England. This happens when soil is no longer protected by the hedges. In this case it is blown away by very strong winds.

Activities

1 Spot the differences in the two sketches in **A**. Try to find one difference for each of the following:
field size, farm buildings, machinery, hedgerows, wetlands.

2 Find a partner in your class. One of you will be a farmer, the other will be a conservationist.

a If you are the farmer, say:
- why you planted hedges in the first place
- why you now wish to remove them.

b If you are the conservationist, explain why you feel it is very important to keep the hedges.
Both of you should get help from photo **C**.

EXTRA

Find out what is meant by **soil erosion**. Why are parts of eastern England affected by soil erosion when it is very windy? Apart from losing it by wind, how else may farmers lose their soil?

Summary

The present appearance of much of Britain's landscape is the result of farming. As farming changes, then so too does our landscape.

How has farming changed?

Farming, like any industry, has to change with the times. It has to keep up with modern methods, provide the produce that people want, look after the countryside, and ensure a living for its workers.

In the last 40 years or so, changes in farming have been more rapid than in the past. Some of these are shown in cartoon **A** below.

The most visible change in farming is that farms are now much bigger. This makes them more efficient. Fields are also larger so that machinery can be used and huge areas given to one crop. The greater use of fertiliser, pesticides and improved seeds has also helped increase the amount of food produced.

However, modern agricultural methods have brought some serious problems. The introduction of machinery has reduced the number of farm workers needed. The use of chemicals has damaged the environment and **wildlife habitats** and **plant communities** have been destroyed.

Efforts have been made to reduce this damage through the introduction of several government initiatives. These include **set-aside** land and **Farm Woodland Schemes**. The **Countryside Stewardship Scheme** is a more recent approach. Through this, the farmers are paid to repair walls, replant hedges, create footpaths and restore damaged habitats. Some of these initiatives are shown on diagram **C** on page 35.

The introduction of these initiatives has had a considerable effect on our countryside. In many areas the landscape is gradually returning to a more natural state and the amount, and diversity, of wildlife is slowly increasing.

A

Farmers

Prices have fallen so we do not make so much money.

Many farmers faced ruin after foot and mouth disease.

The EU is giving us less money for our produce, especially cereals and milk.

New machinery is too expensive for us to afford it.

The rapid increase in fuel prices has cost us a lot of money.

We now earn only just enough money to cover our yearly running costs.

Townspeople

We want farmers to reduce the amount of fertiliser and pesticides they use.

We do not want farmers to grow crops that have been genetically modified.

We would prefer natural, organic crops but do not want to pay high prices.

We want farmers to restore hedges and wetlands for wildlife.

We want access to the countryside and to be able to walk where we want.

Despite the improvements in farming methods and the increase in yields, many farmers still struggle to make a living and many have even gone out of business. Some of the more inventive have developed leisure activities to bring in money and subsidise their incomes. These developments tend to be more successful in tourist areas and near towns rather than in the more remote locations. Cartoon **B** shows some typical activities on these so-called leisure farms.

Activities

1 Look at the graphs in **C**.

a Describe the changes in farming shown by each of the graphs.

b What is the link between the number of tractors and the number of farm workers? Why do you think there is this link?

2 Write two letters to your local newspaper. Each letter should be about half a page in length.

a In the first letter give the views of the farmers who are concerned that producing food and making a living is becoming increasingly difficult.

b In the second letter give the views of townspeople who would like farmers to use methods that are less damaging to the environment.

3 a Why are some farmers developing leisure activities on their farms?

b Draw a star diagram to show some of the leisure activities available on a farm.

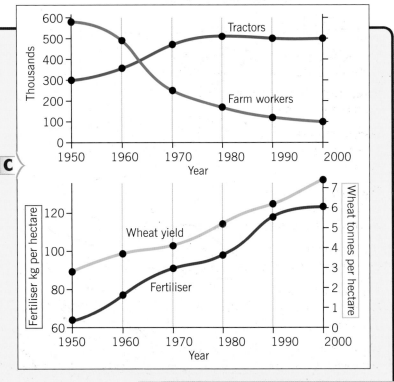

Summary

There have been many changes to farming in recent years. These changes have brought benefits but they have also caused problems.

The farming enquiry

Farmers have to make decisions about what kind of farming they are going to do. They have to choose carefully what crops they will grow or what livestock they may keep. It is important that these decisions are good ones, or the farmer will find it difficult to produce food efficiently and make a living.

In this unit we have seen how the type of farming can be influenced by factors such as climate, relief, soils, transport and government policies. These factors are not the same everywhere but vary from place to place. This helps explain why there are differences in farming types across the country.

In this enquiry, your task is to help four farmers choose the type of farming that is most suited to the farm they have recently purchased. To do this

you will need to look closely at the features of each farm and match these to the requirements of the different kinds of farming.

There should be three main parts to your enquiry.

◆ The first part will be an introduction where you will explain what the enquiry is about.

◆ In the second part you will need to collect and present information to help you make your decisions.

◆ Finally you will need a conclusion. Here you will choose the type of farming for each location and answer the following enquiry question:

How and why does farming vary in different parts of Britain?

Mr McNeish of
Achantee Farm

Mr and Mrs Carr of
Brechie Farm

Mr Tregoglas of
Cutford Farm

Mr and Mrs Nash of
Dalton Farm

A What type of farming?

Farming type	Climate needs	Relief	Soil	Other
Arable Grain crops and vegetables, mainly for human use. Some used as animal feed.	Most crops grow best with: • 650–800 mm of rainfall • summer temperatures above 15°C for ripening and harvesting • sunshine to ripen crops.	Flat or gently sloping so that machinery may be used easily.	Deep and fertile soils that are easy to work.	Near to urban markets and good transport links. Able to use machinery.
Dairying Cattle kept for milk and for making dairy products like cheese and yoghurt.	Good grass needed all year: • more than 800 mm of rain • warm summers • winter temperatures above 6°C for continuous grass growth.	Hilly or gently sloping land. Cows have difficulty on steep ground.	Moisture-retaining for good grass growth.	Near to urban markets and accessible for milk tankers.
Beef cattle Fattened for meat.	Feed mainly on rough pasture. Grass is useful in summer for fattening: • more than 700 mm of rainfall • warm or mild summers.	Hilly or gently sloping land. Cows have difficulty on steep ground.	Can survive on poorer soils if not waterlogged.	Accessible to urban areas for processing. Transport links to markets.
Hill sheep Kept for meat and wool.	Sheep graze on rough pasture. They are hardy animals and can put up with high rainfall (1,500 mm or more), snow, and low winter temperatures.	Sheep are able to graze steep and mountainous land.	Areas of better soil can provide winter feed.	Good access to markets and transport links not so important.

Achantee Farm is in north-west Scotland some 45 minutes by road from Fort William. The roads here are narrow and travel can be difficult. Farmland is mountainous with steep slopes and poor soils. The farm has a small area of flat land on the valley floor close to the river. The soil is deeper here but can be waterlogged in winter.

Brechie Farm is in north-east Scotland on the edge of the Cairngorm Mountains. The land on the farm is hilly but the soils are deep and fertile. There is some good-quality grass on the valley floor, but most of the land is rough pasture. The road are quite good and Aberdeen, the nearest large town, has rail links to Edinburgh and the south.

B

Cutford Farm is in Devon in south-west England. The farm has quite small fields surrounded by high hedgerows with narrow country lanes between them. The countryside is very hilly but the soil is rich and well drained. The M5 motorway is just 5 km away and Exeter, the nearest large town, less than 30 minutes' drive from the farm.

Dalton Farm is close to Cambridge in south-east England. The land here is mostly flat with some gently sloping hills. The farm covers a huge area and has large fields with deep, fertile soil. The M11 motorway runs alongside the farm and there are good rail connections between Cambridge, London and other parts of the country.

The farming enquiry

1 Introduction – what is the enquiry about?

You will need to use maps and writing here. Star diagrams or lists might also help.

a First look carefully at the enquiry question and say what you are going to find out.

- ◆ Describe the main types of farming and their different requirements.

- ◆ Explain how the factors that affect farming vary across the country so that some areas are more suited to certain types of farming than others.

b Describe briefly the main features of the four different farms. Show on a map where they are located.

C Climate conditions

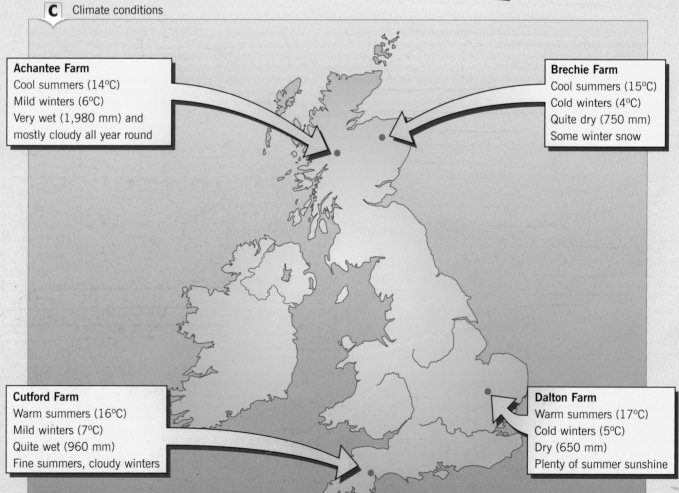

Achantee Farm
Cool summers (14°C)
Mild winters (6°C)
Very wet (1,980 mm) and mostly cloudy all year round

Brechie Farm
Cool summers (15°C)
Cold winters (4°C)
Quite dry (750 mm)
Some winter snow

Cutford Farm
Warm summers (16°C)
Mild winters (7°C)
Quite wet (960 mm)
Fine summers, cloudy winters

Dalton Farm
Warm summers (17°C)
Cold winters (5°C)
Dry (650 mm)
Plenty of summer sunshine

2 What type of farming is most suitable?

a Decide the type of farming most suited to each farm based only on climatic needs. Table **A** and map **C** will help you.

b Now consider other farming needs and check that your choice of farms is still suitable. Table **A** and diagram **B** will help you.

c Make a copy of the farm fact file **D** and complete it to explain your choice of farming. Complete similar fact files for the other farms.

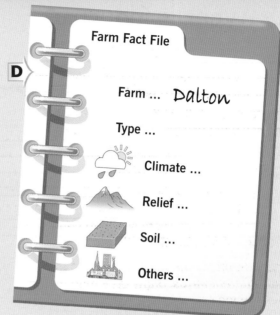

D

Farm Fact File

Farm ... Dalton

Type ...

Climate ...

Relief ...

Soil ...

Others ...

3 Conclusion

Now you should look carefully at your work and answer the enquiry question. There could be two parts to this section.

a First, you could write a short letter to each farmer suggesting what type of farming would be most suited to his farm, and giving reasons for your choice.

b Second, you could describe and suggest reasons for the differences in farming in Britain. This could include writing and a labelled map.

E

What is industry?

What is this unit about?

This unit looks at the different types of work that people do and the factors that affect the location of industry.

In this unit you will learn about:

◆ the differences between primary, secondary and tertiary employment

◆ how to choose the right site for an industry

◆ how the ideal site for an industry may change

◆ the iron and steel industry, the car industry and high-tech industries.

Why is learning about industry important?

Industry is an important part of our lives. It provides things that we need, gives us work so that we can earn a living and helps us achieve high standards of living and a better quality of life.

Learning about industry can also help you:

◆ choose the sort of job that is most suited to you

◆ appreciate why industries have to be located in certain places

◆ understand why some industries have to close down or move away from an area.

A Saltaire, Yorkshire

B

C

Newcastle

- ◆ What might have attracted industry to the place in photo **A**?
- ◆ Which of the workplaces shown in photos **A**, **B** and **C** would you:
  – most like to work in
  – least like to work in?
  Give reasons for your answers.
- ◆ What sort of industry is shown in photo **B**? How is this different from other industries?

What types of industry are there?

Most people have to **work** to provide the things they need in life. Another word for the work they do is **industry**. There are many different types of work and industry. Together they are called **economic activities**. **Economic** means money and wealth.

The work people do can be divided into three main types. These are **primary**, **secondary** and **tertiary**. They are explained below.

A

• **Primary industries** employ people to collect or produce **natural resources** from the land or sea.

• Farming, fishing, forestry and mining are examples of primary industries.

B

• **Secondary industries** employ people to make things. They are usually made from raw materials or involve assembling several parts into a finished product.

• Examples are steel making, house construction and car assembly. **Manufacturing** is another name for this type of industry.

C

• **Tertiary industries** provide a service for people. They give help to others. No goods are made in this type of industry.

• Teachers, nurses, shop assistants and entertainers are examples of people in tertiary industries. This is sometimes called a **service** industry.

In the UK most people now work in either secondary or tertiary occupations. As the graphs in **D** show, however, this has not always been the case.

Before 1800 most people earned a living from the land. The majority were farmers although a few made products that the farmer needed, such as farm equipment, or produced foodstuffs like flour and bread.

During the nineteenth century people turned to industry. An increasing number worked in factories making things like steel, ships and machinery. This was the time of the industrial revolution.

Further changes occurred during the twentieth century and up to the present day. Farming and industry have become more mechanised and need fewer workers. Many people now work in service activities such as in schools, hospitals and shops. Transport has also provided many jobs.

The proportion of people working in primary, secondary and tertiary activities is called the **employment structure**.

UK employment structures **D**

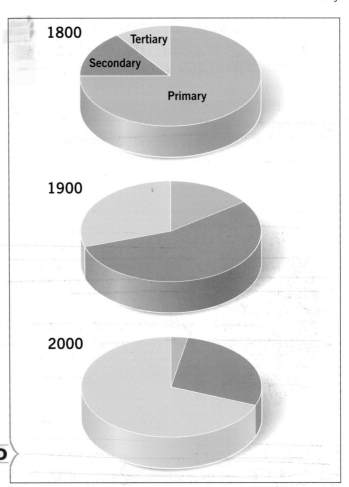

Activities

1 Match the following beginnings with the correct endings.

Primary activities	make things from natural resources
Secondary activities	collect natural resources from the sea or land
Tertiary activities	are grown, found in the sea or dug from the ground
Natural resources	provide a service for people

2 Look at the list of jobs in box **E**. Sort them into primary, secondary and tertiary activities.

3 a Complete a survey of people in your class to find out one job that a member of each person's family does.

b Sort the jobs into primary, secondary and tertiary.

c Draw a bar graph to show your findings.

d Describe what your bar graph shows.

E

Job Shop

- TV presenter
- coal miner
- nurse
- shoemaker
- footballer
- forestry worker
- fire fighter
- bus driver
- carpenter
- fisherman
- police officer
- baker
- shop assistant
- oilrig worker
- quarry worker
- farmer
- pop singer
- builder
- sewing machinist

Summary

The three main types of industry are primary, secondary and tertiary. The proportion of people employed in these industries changes over time.

What is the best site for a factory?

Before building a factory a manufacturer should try to work out the best site for its location. It is unusual to find a perfect site for a factory. Indeed, if there was a perfect site someone else would probably already be using it. Deciding on the best available site depends on several things. Six of these are given in sketch **A**.

A

1 My factory uses lots of **raw materials**. It costs money and takes time to move them so it is best if my factory is as close as possible to these materials. This is even more important if the materials are big and heavy.

2 My factory needs lots of **power** (energy) to work the machines. When the factory was first built a fast flowing river powered the machines. Now we use electricity.

3 A few years ago many people (**labour**) were needed to work in my factory. Today a few machines can do most of this work. However, my present small labour force must be trained and skilled.

4 The best place for any factory is near to a large urban **market**. A market is where most of the people who buy the goods live.

5 **Transport** was a very important factor in choosing the site for my factory. It is needed to bring raw materials and workers to the factory and to send manufactured goods to the market.

6 The **site** for the factory is good because there is plenty of cheap flat land there.

Textiles are often the first industry to be developed in a country. In Britain the first two important textile areas were in East Anglia and the Cotswolds. Later the industry moved to Yorkshire and Lancashire because these areas had more advantages. Yorkshire became important for woollen textiles and Lancashire for cotton textiles. Sketch **B** shows why parts of Yorkshire gave the best location for woollen mills (factories).

B The best location for a textile mill

2 The Pennines get a lot of rain. Rivers flowing from these hills were fast flowing and could power machinery. Later coal was found here and this replaced water power.

1 Raw materials were easy to get. The mills used wool from sheep reared on the Pennines. Water for washing the wool came from local rivers.

3 At this time Britain's population was growing rapidly. Many people needed work and so moved to find jobs in the textile mills.

6 Flat land next to the river made this a good site.

5 Canals were built next to rivers. Canals were used to move both raw materials and manufactured goods. Later railways and roads were built.

4 As Britain's population increased so did the need for clothes. Textiles were needed by people in the growing urban markets both at home and overseas.

Activities

1 Match these beginnings to the correct endings.

Beginnings	Endings
Raw materials are	needed to move raw materials, people and goods.
Power is	a place where manufactured goods are sold.
Labour is	needed to work machines.
A market is	natural resources from which goods are made.
Transport is	people who work in factories.

2 Choose **either** a factory near to where you live **or** the woollen textile mill described on this page. Using diagram **C** as a guide, explain why the factory or mill grew at that site.

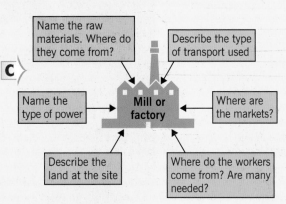

C

- Name the raw materials. Where do they come from?
- Describe the type of transport used
- Name the type of power
- **Mill or factory**
- Where are the markets?
- Describe the land at the site
- Where do the workers come from? Are many needed?

Summary

In choosing the best site to locate a factory, a manufacturer should consider transport and the nearness to raw materials, power sources, workers, and markets for its goods.

13/01/08

Choosing the right site – the iron and steel industry

Industries near to raw materials

Britain was the first country in the world to become **industrialised**. Industrialisation began about 200 years ago after the discovery that coal could be used to produce steam and that steam could be used to work machines. Machines did many of the jobs previously done by people.

In those days transport was poor. There were no lorries or trains, no motorways or railways. Coal and other raw materials were heavy and expensive to move. This meant that most early industries grew up on Britain's coalfields (map **A**). The most important industry became the production of iron and, after 1856, steel.

The iron and steel industry

Three raw materials are needed to make iron and steel – iron ore, coal and, in smaller amounts, limestone. Coke, from coal, is used to **smelt** (melt) the iron ore. This is done because iron ore contains impurities such as carbon. Limestone is added to help separate the pure iron from the impurities, leaving steel behind. Diagram **B** shows what was needed to make one tonne of steel in the nineteenth century.

This meant that early ironworks were located on coalfields where iron ore was found nearby (map **A**).

A Location of iron and coalfields

Key
- Coalfields
- Iron ore
- **S** Early iron and steel areas
- 0   100   200 km

Scottish **S**

North East **S**

Yorkshire, Derby and Nottingham

Western **S**

South Staffs and Midlands **S**

South Wales **S**

Kent

N

B

8 tonnes of coal + 4 tonnes of iron ore + 1 tonne of limestone = 1 tonne of steel

When steel was made instead of iron, steelworks still favoured coalfield locations. Steel was used to make things like ships, trains, bridges and textile machinery. As these industries grew in size and number, many people moved to them to find work. Most coalfields became crowded with people. Today many of these industrial areas have used up their raw materials. Their industries are in decline, factories have closed, it is hard for people to find jobs, and the environment has often been left spoilt.

Pouring molten steel **C**

Britain's largest steelworks is at Port Talbot on the coast of South Wales near to Swansea. Steel has been made here since 1901 when the first steelworks was built. Port Talbot's main product is sheet steel which is used in the car industry, and for making household goods such as washing machines.

So what makes Port Talbot such a good site for a steelworks? Photo **D** shows some of the major reasons. A further important reason is that the government was keen to provide work in an area where many people at the time were out of work and needed jobs.

D | Port Talbot – an ideal site for a steelworks?

Port facilities for import of raw materials and export of steel to foreign markets

Ample supply of water for cooling

Good rail links to UK car factories

Large flat site available on coastal plain

Many skilled workers living in local area

Electric power available through the National Grid

Coking coal and limestone available locally

M4 motorway access to UK markets

Activities

H/W

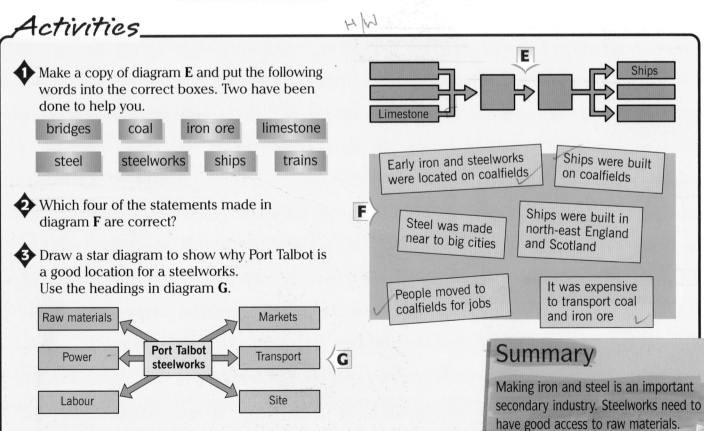

1. Make a copy of diagram **E** and put the following words into the correct boxes. Two have been done to help you.

   bridges   coal   iron ore   limestone

   steel   steelworks   ships   trains

2. Which four of the statements made in diagram **F** are correct?

3. Draw a star diagram to show why Port Talbot is a good location for a steelworks.
   Use the headings in diagram **G**.

E

Limestone → → → Ships

F

Early iron and steelworks were located on coalfields

Ships were built on coalfields

Steel was made near to big cities

Ships were built in north-east England and Scotland

People moved to coalfields for jobs

It was expensive to transport coal and iron ore

G

Raw materials ↔ Port Talbot steelworks ↔ Markets
Power ↔ Port Talbot steelworks ↔ Transport
Labour ↔ Port Talbot steelworks ↔ Site

Summary

Making iron and steel is an important secondary industry. Steelworks need to have good access to raw materials.

13/01/08

How can the ideal site for an industry change?

A factory should try to make a profit. We have already seen on page 48 that before building a factory a manufacturer should work out the best site to locate it. However, the advantages available when locating a textile mill or an iron works in the early 1800s may no longer exist today. This means that the ideal site for a new textile factory or a modern steelworks will have changed.

Changes in the iron and steel industry

Diagram **A** shows a part of Britain which has always been important for making iron and, after 1856, steel. It also shows that while the best sites for the early iron industry were inland, the best location for a modern steelworks is near to the coast.

A

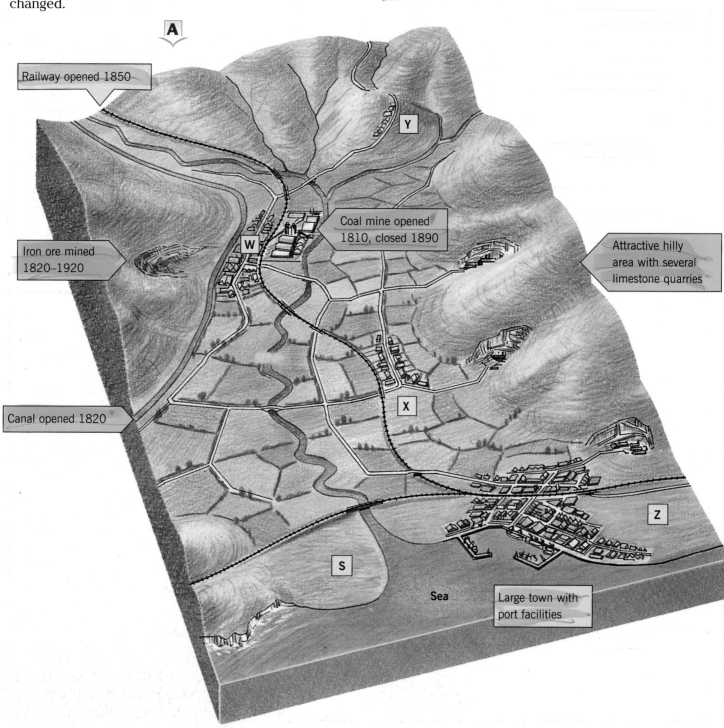

Railway opened 1850

Y

Coal mine opened 1810, closed 1890

W

Attractive hilly area with several limestone quarries

Iron ore mined 1820–1920

Canal opened 1820

X

Z

S

Sea

Large town with port facilities

Activities

H/W

1 In 1820 an iron manufacturer decided to build an ironworks in the area shown on diagram **A**. The manufacturer had to choose between sites **W**, **X**, **Y** and **Z**.

Some of the factors that had to be considered when choosing the site are listed in matrix **C**. The manufacturer eventually chose site **W** for the ironworks. To see why that site was chosen, try completing the matrix yourself.

a Look carefully at diagram **A** and box **B**. For the first location factor give a score for each site. Do the same for each of the other location factors. Part of the matrix has been done to help you.

b Add up the scores. The one with the highest total will be the best site. It should be site **W**. Give four reasons why **W** was a good site for an ironworks in 1820.

2 By 1990 the factory at **W** no longer made a profit. It was too small and the coal and iron ore in the area had run out. These raw materials were now brought in from abroad by sea. It was decided that the factory had to be enlarged or a new steelworks built either at site **Z** or **S**.

Matrix **D** shows some of the factors that the manufacturer had to consider before deciding whether to stay at **W** or go to **Z** or **S**. You decide what should be done by completing the matrix. Remember that it is much cheaper to expand at the same place than to build a new factory at a different site.

a Complete the matrix using the scoring from box **B**.

b Did you decide to expand at **W** or did you choose one of the sites **Z** or **S**?

c Write a short paragraph giving the reasons for your decision.

B

Give a score of 0 to 4 for each site.

4 if the site is **excellent**

3 if the site is **very good**

2 if the site is **good** but has faults

1 if the site is **poor** and only just acceptable

0 if the site is **unsatisfactory**

C

Location factors for ironworks in 1820	Site W	Site X	Site Y	Site Z
Near to local iron ore	4			1
Near to local coal		2		
Near to local limestone				
River or canal needed for transport	4			
Flat land needed for factory				4
Close to town for workers				
Total				

D

Location factors for steelworks in 1990s	Site W	Site Z	Site S
Near to port for coal and iron imports			
Near to local limestone			
Railways and roads needed for transport			
Flat, cheap land needed for factory			
Near town for skilled workers			
Same or different location			
Total			

Summary

The best site for an industry changes as location factors change. The British iron industry grew up near raw materials. Britain's few remaining steelworks are located near to the coast.

Choosing the right site – the car industry

Industries near to markets

As raw materials are used up, and as transport improves, then modern factories tend to locate in areas where many people live. This is mainly because present-day industries need large markets in which to sell their goods. The car industry is an example of an industry that builds new factories near to markets.

The car industry

A modern car consists of many small parts. Each part is made in its own factory. If the factories making these parts are all close together then it is easier and cheaper for the car manufacturer to **assemble** (put together) all of these parts. If large towns are nearby then workers from these towns can make and assemble the parts and, hopefully, buy many of the finished cars. Transport is important for moving car parts, assembled cars and workers. Map **A** shows where most people in Britain live and where the largest car assembly plants were located in 2001. Some of these have now closed or changed ownership as the industry moves on.

Today industrial growth is more likely in those areas where there are most people. In these places new factories are opening, jobs are easier to get, and more care is taken of the environment.

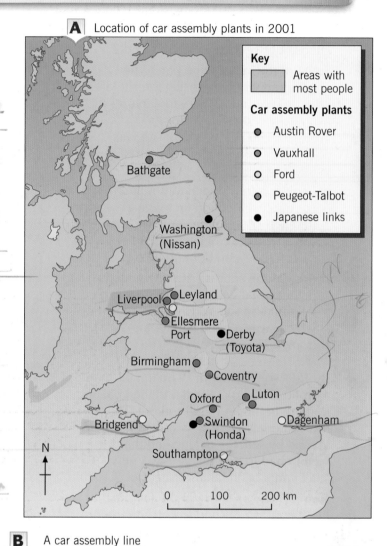

A Location of car assembly plants in 2001

Key

Areas with most people

Car assembly plants
- Austin Rover
- Vauxhall
- Ford
- Peugeot-Talbot
- Japanese links

Bathgate
Washington (Nissan)
Liverpool · Leyland
Ellesmere Port
Derby (Toyota)
Birmingham
Coventry
Oxford · Luton
Bridgend
Swindon (Honda) · Dagenham
Southampton

N

0   100   200 km

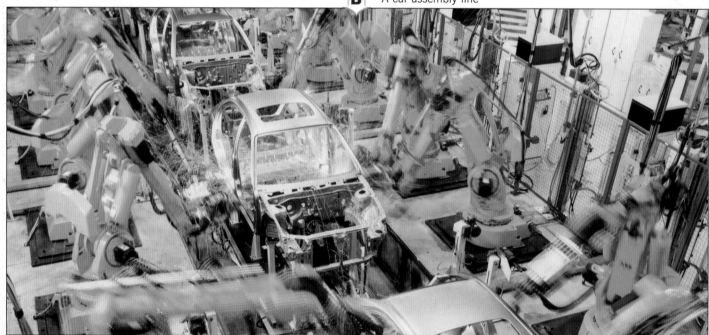

B A car assembly line

Toyota is Japan's largest car company. In the early 1990s it decided to build a new car manufacturing plant at Burnaston near Derby. The plant opened in 1992 and in 2005 was producing 220,000 cars a year with a workforce of 3,800 people. Most Toyota cars are sold in the UK and Europe. Some are even transported to Japan to be sold there.

Toyota uses a **just-in-time** system of manufacture where components (car parts) are supplied to the assembly line just minutes before they are needed. Expensive parts do not have to be stored on site so costs are reduced. Just-in-time needs a good transport system for it to work. Map **C** shows some reasons why Toyota chose Burnaston.

C Burnaston – an ideal site for a car manufacturer?

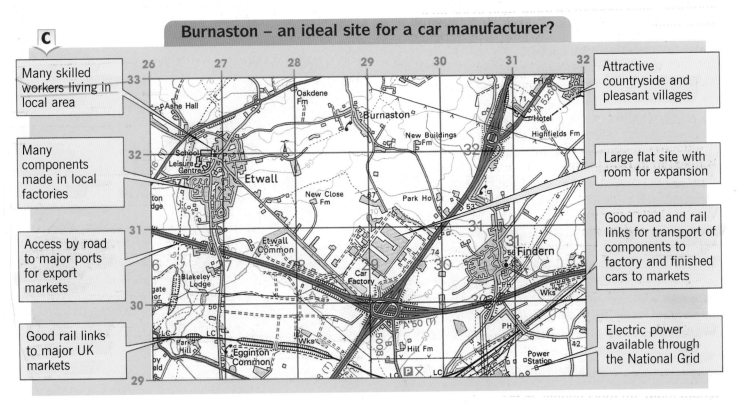

Many skilled workers living in local area

Many components made in local factories

Access by road to major ports for export markets

Good rail links to major UK markets

Attractive countryside and pleasant villages

Large flat site with room for expansion

Good road and rail links for transport of components to factory and finished cars to markets

Electric power available through the National Grid

Activities

1. Of the six statements shown in drawing **D**, four are correct. Write out the correct ones.

2. Car factories are usually located close to large towns. Give at least two reasons for this.

3. What are the advantages and disadvantages of the just-in-time system?

4. Explain why Burnaston is a good site for a car factory. Use the headings in fact file **E**.

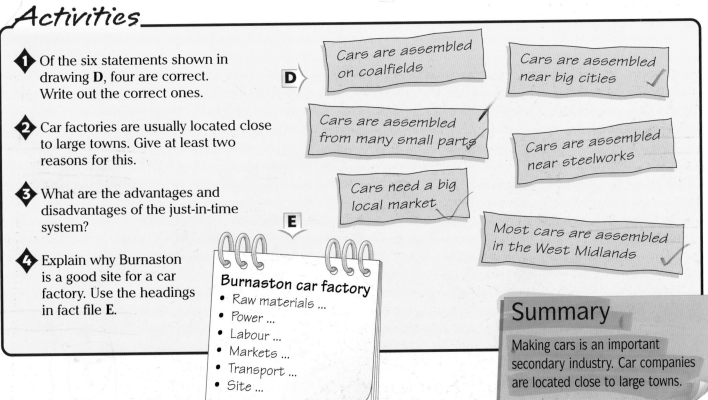

D >

Cars are assembled on coalfields

Cars are assembled near big cities ✓

Cars are assembled from many small parts

Cars are assembled near steelworks

Cars need a big local market

Most cars are assembled in the West Midlands ✓

E >

Burnaston car factory
- Raw materials ...
- Power ...
- Labour ...
- Markets ...
- Transport ...
- Site ...

Summary

Making cars is an important secondary industry. Car companies are located close to large towns.

What are high-tech industries?

High-technology or **high-tech industries** make products such as microchips, computers, mobile phones, pharmaceuticals (drugs) and scientific equipment. They have been the growth industry of recent years and now provide more than 25 per cent of the UK's manufacturing jobs.

High-tech companies use the most advanced manufacturing methods. They put great emphasis on the research and development of new products and employ a highly skilled and inventive workforce. Most are huge organisations with offices and factories throughout the world. The UK electronics industry, for example, is controlled almost entirely by foreign companies, mainly from Japan and the USA.

Firms that make high-tech products often group together on pleasant, newly developed **science** or **business parks**. All firms on a science park are high-tech and have direct links with a university. Business parks do not have links with universities and may include superstores, hotels and leisure centres. There are many more business parks than there are science parks.

Both are located on edge-of-city **greenfield sites** although some business parks are found in inner city areas that have been redeveloped, such as London Docklands. Photo **B** shows the sort of building to be found on a science park.

A Some high-tech companies

B

Mature trees to screen buildings

Buildings have plenty of space

Large car parking areas

Modern buildings with central heating, air conditioning and large windows for light

Grassy areas, ornamental gardens, lakes and ponds

Diagram **C** shows some of the advantages to industry of science and business parks. Others include:

◆ They should be near to a main road on the edge of town for easy access.

◆ Nearby firms can exchange ideas and information.

◆ Leisure facilities and support services may be shared.

◆ A pool of highly skilled workers can be built up.

There can, however, be some disadvantages. For example:

◆ An over-use of cars can cause traffic congestion at busy times.

◆ Edge-of-town sites can be far from shops and services in the town centre.

◆ Firms may prefer to be by themselves so as to keep new ideas a secret.

◆ At times it may be difficult for firms to find enough skilled workers.

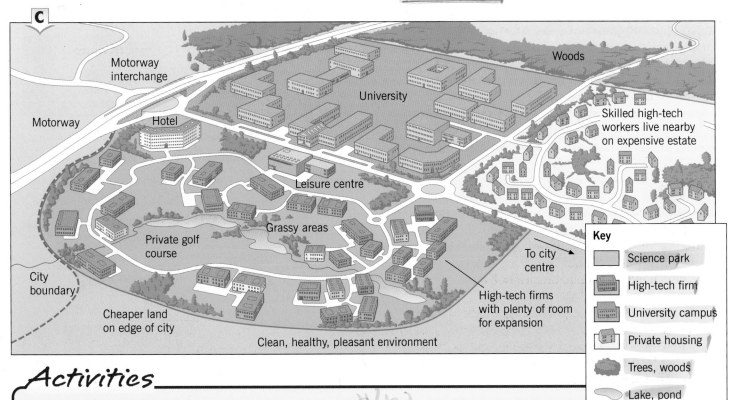

C

Motorway interchange

Motorway

Hotel

University

Woods

Skilled high-tech workers live nearby on expensive estate

Leisure centre

Grassy areas

Private golf course

City boundary

To city centre

High-tech firms with plenty of room for expansion

Cheaper land on edge of city

Clean, healthy, pleasant environment

Key

	Science park
	High-tech firm
	University campus
	Private housing
	Trees, woods
	Lake, pond

Activities

1
  a What are high-tech industries?

  b Name five high-tech products that you have used in the last week.

2 Complete table **D** to show the differences between a science park and a business park. Choose your answers from the following pairs:

  a many/very few

  b university links/no university links

  c high-tech firms/high-tech firms, shops, hotels and leisure centres.

D

	Science park	Business park
a		
b		
c		

3 Using photo **B** and diagram **C**, give six reasons why a high-tech firm should locate on a science park. In your answer you should mention each of the following:

transport, price of land, the environment, people's health, leisure facilities, exchanging ideas with people from other firms

4 Give two disadvantages which may arise from firms locating in the same place.

Summary

High-tech industries use advanced scientific techniques. They often locate on edge-of-town science or business parks.

Where are high-tech industries located?

Industries like shipbuilding, steelmaking, chemical manufacture and textiles used to be very important in the UK. They employed hundreds of thousands of people, mainly in the north and west of the country. These industries are now in decline and large numbers of jobs have been lost as companies have closed down or reduced their output. Industries in decline are often called **sunset industries**.

Sunrise industries are growth industries. They include high-tech industries which use modern factories and often have their own research and development units. High-tech industries have a much freer choice of location than the old traditional industries where nearness to market and raw materials were so important.

The list in **A** shows some of the factors that have to be considered when choosing the location for these industries. A typical site for a high-tech factory is shown in photo **B** which is on a greenfield site, on the edge of a country town, near to a motorway junction and close to a university.

A

High-tech industries like to be located:

◆ near to motorways or good roads
◆ near to highly qualified and skilled workers
◆ near to research facilities in universities
◆ near to pleasant housing and open space
◆ near to attractive countryside and good leisure facilities
◆ near to an airport for international links.

B

Although high-tech companies have factories and research centres in many parts of Britain, three main areas have become particularly important. These are:

1 The M4 corridor following the motorway westwards from London ('Silicon Strip')

2 'Silicon Glen' in central Scotland

3 'Silicon Fen' in and around Cambridge.

These locations are shown on maps **C** and **D**.

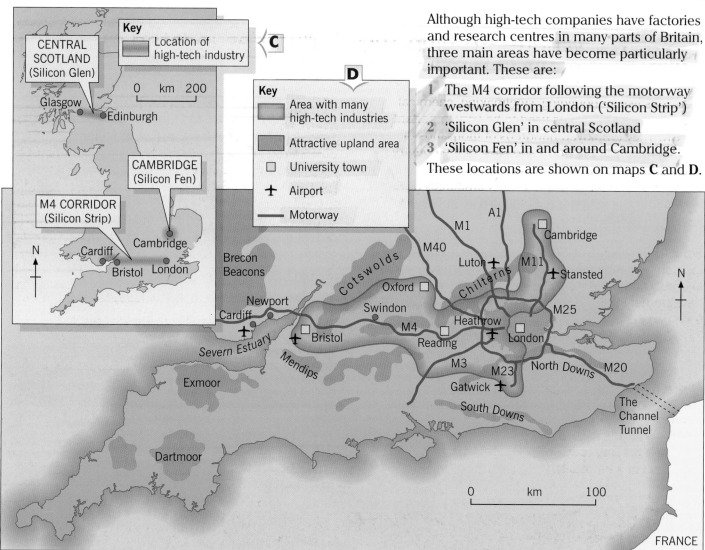

C

Key
Location of high-tech industry

CENTRAL SCOTLAND (Silicon Glen)

Glasgow · Edinburgh

0   km   200

CAMBRIDGE (Silicon Fen)

M4 CORRIDOR (Silicon Strip)

Cardiff · Bristol · London · Cambridge

N

D

Key
Area with many high-tech industries
Attractive upland area
□ University town
✚ Airport
── Motorway

Cambridge · A1 · M1 · M40 · Luton · M11 · Stansted · Brecon Beacons · Cotswolds · Chilterns · Oxford · Swindon · Heathrow · M25 · Newport · Cardiff · Bristol · M4 · Reading · London · Severn Estuary · Mendips · M3 · M23 · North Downs · M20 · Exmoor · Gatwick · South Downs · The Channel Tunnel · Dartmoor

N

0   km   100

FRANCE

Activities

1 What are the differences between sunset and sunrise industries?

2 Drawing **E** is a sketch of photo **B**.

a Make a larger copy of the sketch.

b Colour lightly in pencil: the housing area red, the industrial area brown, the main roads yellow and the countryside green.

c Label the features that make the area a good site for a high-tech factory. Write about 10–20 words of explanation for each one.

3 Use the information on map **D** to explain why the M4 corridor is a good location for high-tech industries.

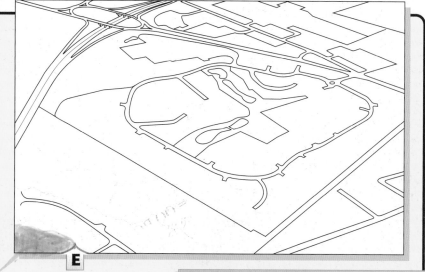

E

Summary

High-tech industries can locate in a wide variety of places. Most are found in central Scotland and the south and east of England.

Industrial location enquiry

This enquiry is concerned with locating a factory. You will find pages 48 to 55 of this book helpful as you work through the enquiry.

In recent years a number of major car manufacturers, such as Nissan, Honda and Toyota, have set up new factories in different parts of the UK. All are now producing cars that are made in the UK and sold throughout Europe and across the world.

Setting up these factories took many years of planning and preparation. One of the most important decisions that had to be made in the planning process was where to locate the factory. In most cases the responsibility of choosing the best site was given to a small team of experts.

The team began by making a list of location factors for the new factory. They then drew up a shortlist of sites that met the basic requirements.

Next, they investigated the advantages and disadvantages of each site before finally choosing what they considered to be the best site.

Some of the location factors that the team considered are shown in the drawing below. On the opposite page are ten possible sites along with information about each one.

In this enquiry you should imagine that you were one of the experts working with the team. Your task is to suggest:

◆ the best site for a car factory
◆ how you came to make that choice
◆ whether their choice was the best one.

Where is the best site for a new car factory?

A

Good **communications** are very important to us. We need to be close to a motorway and quite near to a port or airport.

We require a very large **site**. To start with we must have at least 2 km². We will need more later for expansion. The land needs to be flat and well drained.

We need a good **workforce**. For this there must be a large number of workers living nearby. There should be good relations between the workforce and the company.

We would like to be located in a pleasant **environment** where our workers can enjoy good living conditions and be near to attractive countryside.

We must have **government aid**. This will help us finance the project, give us support in setting up the factory and provide training for our new workforce.

Possible sites for a new car factory

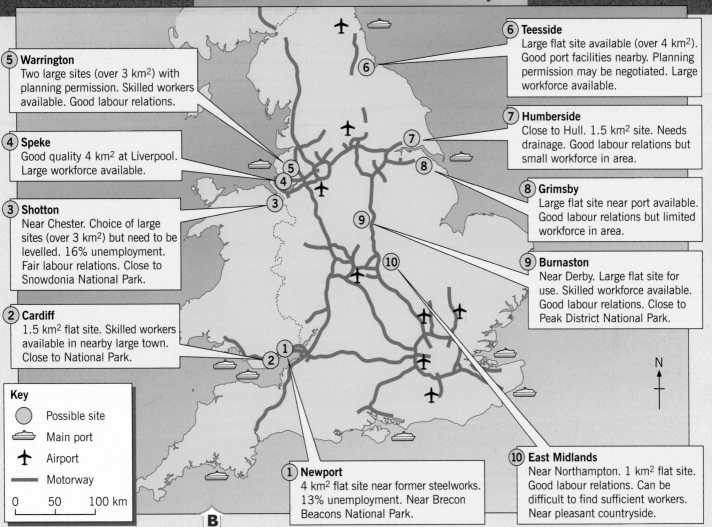

5 Warrington
Two large sites (over 3 km²) with planning permission. Skilled workers available. Good labour relations.

4 Speke
Good quality 4 km² at Liverpool. Large workforce available.

3 Shotton
Near Chester. Choice of large sites (over 3 km²) but need to be levelled. 16% unemployment. Fair labour relations. Close to Snowdonia National Park.

2 Cardiff
1.5 km² flat site. Skilled workers available in nearby large town. Close to National Park.

6 Teesside
Large flat site available (over 4 km²). Good port facilities nearby. Planning permission may be negotiated. Large workforce available.

7 Humberside
Close to Hull. 1.5 km² site. Needs drainage. Good labour relations but small workforce in area.

8 Grimsby
Large flat site near port available. Good labour relations but limited workforce in area.

9 Burnaston
Near Derby. Large flat site for use. Skilled workforce available. Good labour relations. Close to Peak District National Park.

1 Newport
4 km² flat site near former steelworks. 13% unemployment. Near Brecon Beacons National Park.

10 East Midlands
Near Northampton. 1 km² flat site. Good labour relations. Can be difficult to find sufficient workers. Near pleasant countryside.

Key
- ⬤ Possible site
- ⛴ Main port
- ✈ Airport
- ▬ Motorway

0   50   100 km

N

B

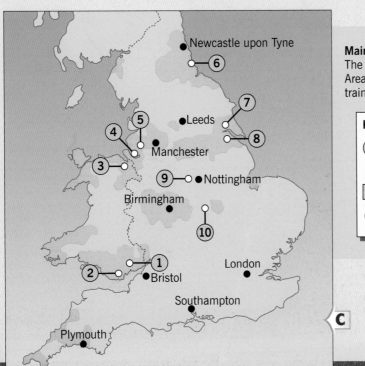

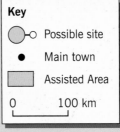

Newcastle upon Tyne
Leeds
Manchester
Nottingham
Birmingham
Bristol
London
Southampton
Plymouth

Main towns and Assisted Areas
The government gives help to industry in Assisted Areas. This may include loans, grants, help with training and provision of services.

Key
- ⬤○ Possible site
- ● Main town
- ▭ Assisted Area

0   100 km

C

Industrial location enquiry

1 Introduction – what is the enquiry about?

You could use diagrams, maps, writing or lists here.

a Describe what you have to find out in this enquiry.

b Explain how the best site may be chosen.

c Describe the possible sites.

2 What are the location factors?

Draw a star diagram like the one here. Add about ten words of description to each factor. The diagram on page 60 will help you.

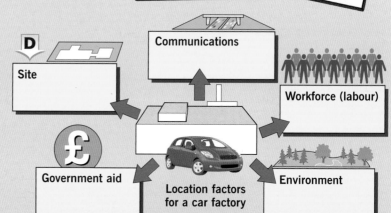

Matrix for choosing the best site for a new car factory

E

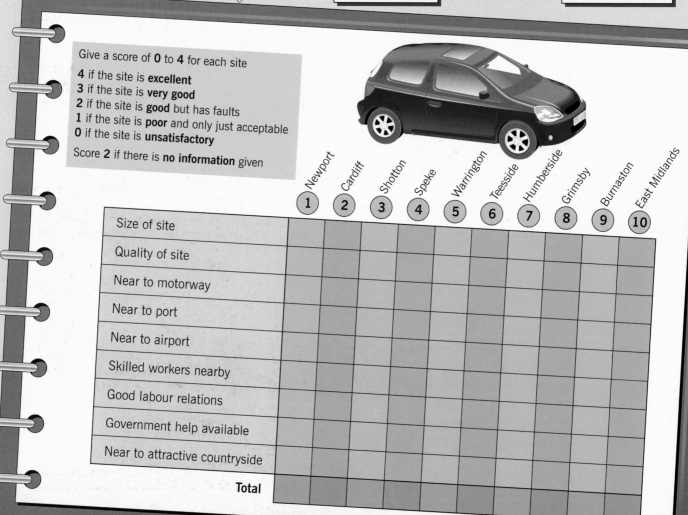

Give a score of **0** to **4** for each site

4 if the site is **excellent**
3 if the site is **very good**
2 if the site is **good** but has faults
1 if the site is **poor** and only just acceptable
0 if the site is **unsatisfactory**

Score **2** if there is **no information** given

	Newport (1)	Cardiff (2)	Shotton (3)	Speke (4)	Warrington (5)	Teesside (6)	Humberside (7)	Grimsby (8)	Burnaston (9)	East Midlands (10)
Size of site										
Quality of site										
Near to motorway										
Near to port										
Near to airport										
Skilled workers nearby										
Good labour relations										
Government help available										
Near to attractive countryside										
Total										

3 Which is the best site?

a Make a larger copy of the matrix opposite.

b Look carefully at the information on page 61. For the first location factor give a score for each site. Do the same for each of the other location factors. (The location factors are described more fully in the diagram on page 60.)

c Add up the scores in each column. The one with the highest total will be the best site.

d List the ten sites in order of their total scores. Put the best at the top of the list.

4 Conclusion

a Which are the three best sites? List the main advantage of each one.

b Which of the three would you choose? Suggest why this site is better than the other two.

c If your preferred site had not been available, would the other sites have been satisfactory? Give reasons for your answer.

d Either
   ◆ Draw a simple sketch of the photo below. Add information to the boxes to explain why your choice is a good site for a car factory. Page 61 and your completed matrix will help you with this.

   Or
   ◆ Write a report for your Head Office about your chosen site. In the report describe the main features of the site and explain why it is a good location for a car factory.

Workforce (labour)

Site

Government aid

Communications

Environment

F

What is the environment problem?

What is this unit about?

This unit looks at how the environment may be damaged by the misuse of resources and shows how it may be protected by careful management.

In this unit you will learn about:

◆ who cares for the environment
◆ how environments can be damaged
◆ the need to protect wildlife and scenery
◆ renewable and non-renewable resources
◆ the problems of oil and electricity production
◆ how to conserve resources and protect the environment.

Why is this environment topic important?

The environment is where we live, work and spend our leisure time. It includes resources which we need to ensure a good quality of life. It is important for us and our children that we learn how to use resources without damaging the environment.

Learning about the environment can help you:

◆ understand how we use the earth's resources
◆ recognise the need to recycle waste materials and reduce energy consumption
◆ appreciate the need to protect and conserve wildlife and scenery
◆ develop an interest in your surroundings.

A Hemel Hempstead, 2005

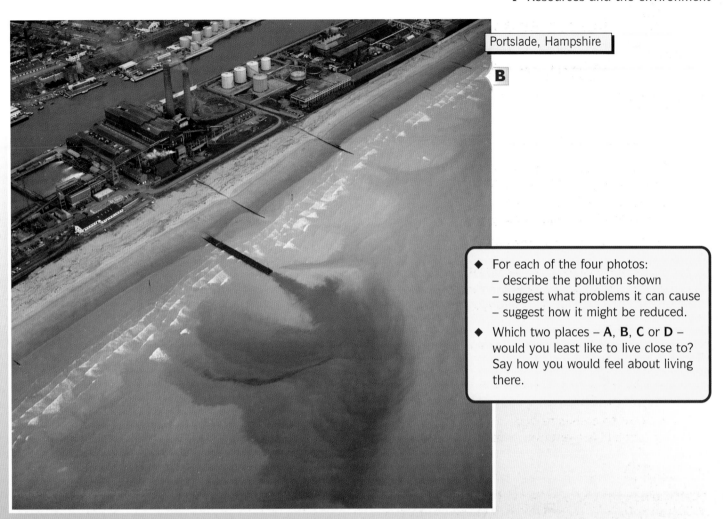

Portslade, Hampshire

B

◆ For each of the four photos:
 – describe the pollution shown
 – suggest what problems it can cause
 – suggest how it might be reduced.

◆ Which two places – **A**, **B**, **C** or **D** –
 would you least like to live close to?
 Say how you would feel about living
 there.

C Ribblesdale, Yorkshire Dales

D Heathrow, London

Why are we concerned about the environment?

Conservation of resources needs careful **planning and management**. Planning allows people to use resources to their best advantage without polluting the environment. Planning and management reduces conflict between groups of people and between people and the environment.

Planning means making decisions where there is a conflict of interest. For example, how can we build houses, create jobs and develop leisure activities while at the same time protecting the environment.

The **environment** is everything around us. It includes the natural physical surroundings where people, plants and animals live.

Conservation of resources is very important. Conservation is the protection of resources for future use and limiting damage to the environment. This is known as **sustainable development**.

It is important to try and conserve the air, the sea, minerals, energy, soil, trees, wildlife and scenery.

up!

Activities

1 **a** Put these boxes in the correct order to explain why it is important to plan and manage the environment.
b Give an example for each box.

Resources are natural materials which are useful to people. 2

Pollution happens when people harm and destroy the environment. 3

Conservation is when people use resources carefully and protect the environment. 4

Planning and management is when people use resources without spoiling the environment. 5

The **environment** is the natural surroundings where people and animals live. 1

The environment contains **resources** which are materials that occur naturally. They are useful because they can improve people's lives.

Coal, iron ore, clean water, soil, trees, mountains, wildlife and sandy beaches are all examples of resources.

People must use resources carefully. A careless use may mean that some resources will be **used up**, some will be **wasted**, and some may cause **pollution**. Pollution is when people harm or destroy the natural environment. Pollution can kill plants and animals.

People often misuse these resources. Some resources, like coal, are used up. Some, like forests and soil, are destroyed. Others, like rivers and the air, become polluted.

down!

2 Think about one normal day in your life. During the day you probably used up several resources, caused quite a lot of pollution and may have tried to conserve the environment. Draw and complete a table like the one below to show, 'One day in my life'. Some ideas have been added to the table to help you.

Resources which I used	Pollution which I caused	How I tried to conserve the environment
• Clean water	• Dirty water after my shower	• Went to the bottle-bank
• Had fish and chips at lunchtime	• Left chip paper in school playground	• Put empty drinks can in dustbin

EXTRA

By yourself, as a group, or as a class, think of one way in which you can try to conserve and protect the environment. Try it out for one week. After the week:

- decide if you were successful or not
- discuss with others why it is often hard to protect the environment.

Summary

If resources are misused they can be used up or they may harm the environment. It is important that resources are used sensibly and the environment is protected for the future. This needs careful planning and management.

Who cares for the environment?

One of the jobs of the British government is to try and protect the countryside. Three government bodies are:

◆ the Countryside Agency which looks after country parks and long distance footpaths

◆ the Forestry Commission which is mainly interested in coniferous forests

◆ the Nature Conservancy Council which has developed Nature Reserves.

However, there are also many voluntary groups of people (**conservationists**) who are concerned with protecting the environment. Some of these groups are named in diagram **A**. All of these groups have to rely upon donations of money to help them do their job.

A

Each group tends to concentrate upon one, and sometimes more, aspects of the environment. Some look after areas of **attractive countryside**, some look after **wildlife habitats** while others are more interested in **historic sites** (diagram **B**). They are all working to stop further damage to the environment.

B

Areas of attractive countryside

◆ Rivers   ◆ Coasts
◆ Lakes   ◆ Moors
◆ Mountains

Wildlife habitats
(homes of plants and animals)

◆ Woods
◆ Hedges
◆ Wetlands (marshes)

Historic sites

◆ Stately homes
◆ Castles
◆ Industrial museums

Activities

1 What do we mean by the following terms?

| conservationist | attractive countryside |

| wildlife habitat | historic site |

2 Sketch **C** shows several different environments which need special protection. These environments are areas of attractive countryside, wildlife habitats and historic sites. They have been numbered **1** to **12**.

a Which three are likely to be looked after by government bodies?

b Which of the voluntary groups named in diagram **A** may look after each of the remaining nine environments?

C

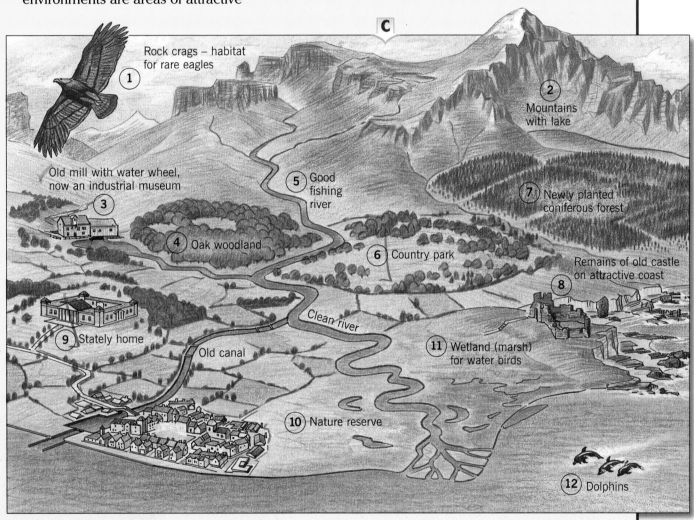

1 Rock crags – habitat for rare eagles
2 Mountains with lake
3 Old mill with water wheel, now an industrial museum
4 Oak woodland
5 Good fishing river
6 Country park
7 Newly planted coniferous forest
8 Remains of old castle on attractive coast
9 Stately home
Clean river
Old canal
10 Nature reserve
11 Wetland (marsh) for water birds
12 Dolphins

EXTRA

People in the town on sketch **C** like visiting the mountains and lake. They need a better road to make their journey easier and faster. Where would you build the road so that it did not upset any conservation group? Describe your route and say what problems there might be.

Summary

Some types of environment need special protection. These environments include areas of attractive countryside, wildlife habitats and historic sites. Some of these environments are looked after by government bodies, while others are protected by voluntary conservation groups.

Why does wildlife need protecting?

Wildlife is always under threat from people. Some **species** (types) of plants, animals, birds and fish have already become **extinct**. Extinct means that there are no more of that species still living. Many other species of wildlife are said to be **endangered**. Unless something is done quickly to help and protect these endangered species they too will become extinct. Table **A** and the photos in **B** give examples of some endangered species, together with reasons why they have become endangered.

A

Examples of endangered species	Reasons for becoming endangered
Crocodiles and alligators	Skin is used for shoes, handbags and belts
Leopards, cheetahs and jaguars	Fur is used for clothes
Hawksbill turtles	Can no longer breed on beaches since these are full of tourists
Rhinos	Killed for their horns. A dagger with a rhino horn handle costs £10,000
Blue butterflies	Virtually extinct in the UK as their natural habitat has been destroyed
Whales and dolphins	Killed for food, caught in fishing nets
Elephants	Killed for their ivory
Other endangered species include gorillas, giant pandas, tigers, parrots, orchids and many cacti plants.	

B

Elephants in Kenya – the problem

The long tusks of elephants are made of ivory. Ivory is taken from dead elephants and many have been killed for their tusks. Their trunks are often cut off to make the job of hacking out the tusks with axes easier. Most of the ivory from Kenya was sent to places like Hong Kong and Japan to be made into ornaments. As the world's trade and the price of ivory increased, more and more elephants were killed. Even when the Kenyan government created the country's first National Park in 1948, it did not stop the many poachers from killing elephants. Of the 111,000 elephants in Kenya in 1973, only 65,000 were left by 1981 and 19,000 in 1987. In fourteen years more than eight out of every ten elephants were killed. The elephant had become an endangered species.

Kenya needs its elephants. They, and other animals, are an important source of wealth to the country because they attract tourists (photo **C**). Many people come from overseas to go on safari, which means they can see wildlife in its natural habitat. Elephants are also useful to other wildlife in the area. They dig for water in dry areas, making it available to other species, they make tracks for smaller animals through thick vegetation and, on death, provide food for many predators for many weeks.

What has been done?

♦ Conservation groups like WWF (World Wide Fund for Nature) raised money to help protect the elephant. They helped set up a conservation group called CITES (Convention on International Trade in Endangered Species). This led in 1990 to a world ban on trading in ivory.

♦ Education within Kenyan schools, asking people in rich countries to sponsor an elephant, and advertising on T-shirts, have all helped protect the elephant (cartoon **D**).

By the mid-1990s, the number of elephants, especially in southern African countries, had begun to increase. These countries suggested that elephants can be treated as a 'crop', and surplus elephants could be 'harvested' under strict control. Money from the sale of ivory, skins and meat could then be used to help local communities and pay for conservation. Since then the CITES ban has been lifted in several countries.

C

D

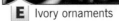

E Ivory ornaments

Activities

1 a What is meant by 'endangered species'?
b Give **five** reasons why some species have become endangered.

2 Design **either** a poster **or** a T-shirt to remind people that ivory souvenirs will have cost an elephant its life.

3 Choose one endangered species. It might be one shown in the photos in **B** or another one that you know about. Explain why that species has become endangered. Give reasons why you think that it needs to be protected.

Summary

Many species of wildlife have become rare and are threatened with extinction. They need careful protection if they are to survive.

ivory is more beautifap on elephants than people

How can industry pollute the environment?

Industry has helped to make Britain one of the richer countries in the world. It makes goods that have improved our standard of living. But industry uses up **resources**, creates waste and can damage the environment.

In the past the main aim of industry was to make a profit. There was little concern or interest in the environment, especially as any damage caused by factories often affected only the local area. People in the main industrial towns of Britain just accepted smoke-filled skies, smoke-stained buildings, dirty rivers with no fish, and piles of coal and other industrial waste, as a normal part of their environment.

Today the effects of industry are so widespread that they are threatening the whole global environment.

Types of industrial pollution

Air

Smoke from chimneys can affect human health by causing breathing problems. Carbon dioxide in smoke is causing world temperatures to rise (see page 127). Other gases in smoke cause acid rain which kills trees and fish.

Water

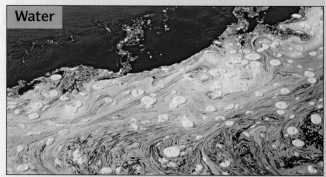

Factories can allow waste materials to escape into rivers. Industries use water for cooling and return warm water to rivers. This can kill fish.

Visual

Few factories are attractive to look at. When factories close, the land may be left empty (derelict).

Some industries dump their waste materials. Sometimes this waste is dangerous, poisonous or radioactive.

Noise

Lorries going to and from factories, and heavy machinery, create noise for people living nearby.

Smell

Some industries, such as the plastics industry, give off very unpleasant smells.

What can be done to reduce industrial pollution?

One way is by education. As more people become aware of the dangers of pollution, more pressure can be put on industry and governments to prevent further pollution. Results in the last few years have been more encouraging but it is impossible to solve all the problems at once. Likewise, it is impossible to clean up a polluted environment overnight.

A second way is to spend more money on preventing pollution. Unfortunately, not all industries and governments have the money, or the desire, to spend what little money they have on stopping pollution. Factories, and power stations producing energy for industry, can reduce the amount of smoke and gases they give out (page 79). However, this may mean a drop in profits or an increase in the price of goods.

Certainly there is an urgent need to plan and manage our environment more carefully. Prevention is easier than cure.

Activities

1 Look at cartoon **B**.

a Name the five types of pollution shown on the cartoon.

b What two types of pollution would be the most damaging to plants and wildlife? Give reasons for your answer.

c Imagine that you lived or worked in this area. Which two types of pollution would you dislike the most? Give reasons.

2 a How can education help to reduce pollution?

b Why do many industries not want to spend money on trying to reduce pollution?

3 When applied to pollution, what is meant by the term, 'prevention is easier than the cure'?

EXTRA

Work with a partner and produce a poster to show how industry pollutes the environment. Add a slogan to draw people's attention to these dangers.

Summary

Industry contributes towards air, water, noise, visual and smell pollution. Much education and money are needed to prevent pollution in the future. Cleaning up environments is costly, difficult, and may take a long time.

17/3/09

How can environments be damaged?

The mining of coal, the quarrying of slate and the collection of gravel are important **primary industries**. They provide the **raw materials** that we need, are a source of employment for local people and earn money for the country. However, the collection of raw materials, whilst bringing many benefits, can also cause problems and be harmful to the environment.

Photo **A** shows a quarry in a **National Park**. The rock that is being extracted is used in the building industry and for roads. There are 30 working quarries in our National Parks extracting 12 million tonnes of rock every year. The quarries are a major source of conflict as they tend to destroy the very landscape that the National Parks are supposed to protect. Photo **A** shows some of the causes of this conflict.

A

This is even done in Sri Lanka

Problems

◆ Quarries are dirty and dangerous.
◆ Noise is caused by blasting and heavy lorries.
◆ Dust from blasting and lorries causes air pollution.
◆ Wildlife is frightened away.
◆ Buildings and spoil heaps look ugly.
◆ Heavy lorries cause traffic congestion.

Benefits

◆ Quarries provide work and income for local people.
◆ A quarry is a source of money for the local council.
◆ Local roads are improved for the increase in traffic.
◆ Quarries provide important raw materials for the nation.
◆ Slate and limestone can be used locally.
◆ Primary industries can attract other work to the area.

I am sleepy

Activities

make the that makes 2 of us.

1 Make a larger copy of table **B** below. Complete the table with the help of photo **A**. Your completed table should show how quarrying can harm the landscape and affect people and wildlife.

B

Harmful effects of quarrying on:		
Landscape	People	Wildlife

2 Imagine that you live close to the quarry in photo **A**. List the benefits that the quarry could bring to you and your family.

3 Look at photo **C** and drawing **D**. In 1978 a 'Warning – Keep Out' notice was put up in the quarry. Draw the notice and add what you think might have been written on it.

How can damaged environments be improved?

National Park authorities try to ensure that quarries are landscaped and screened when they are working. They also insist that when a quarry closes, it should be restored to its original appearance.

Drawing **D** shows a restored quarry near the Roman Wall in Northumberland National Park. The quarry had been used for over 50 years and eventually closed in 1978. When it was closed it left a hole 30 metres deep and the size of 16 football pitches. The hole was flooded and was dangerous. The quarry also had many old, unused buildings and piles of mined waste.

Some of the ways in which the quarry has been restored and the environment improved since closure are shown in drawing **D**. Photo **C** shows how far the improvements had reached by 2005. It is now a popular leisure attraction.

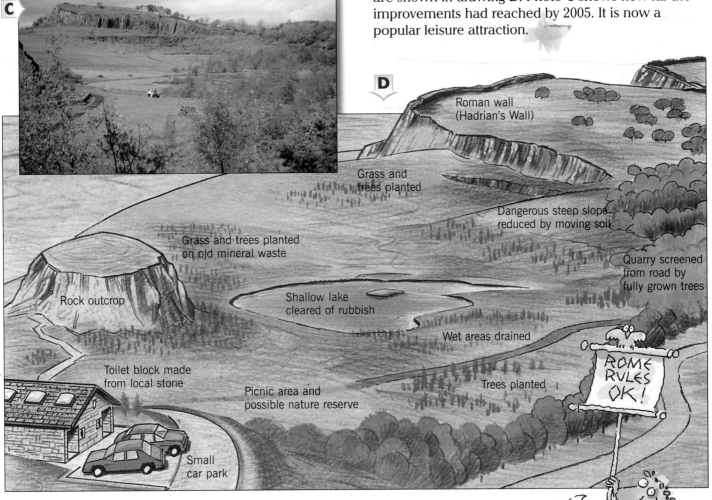

C

D

Roman wall (Hadrian's Wall)

Grass and trees planted

Dangerous steep slope reduced by moving soil

Grass and trees planted on old mineral waste

Quarry screened from road by fully grown trees

Rock outcrop

Shallow lake cleared of rubbish

Wet areas drained

Toilet block made from local stone

Trees planted

Picnic area and possible nature reserve

ROME RULES OK!

Small car park

It's changed a lot since I was last here!

4 How have the changes made to the quarry improved the landscape for:

a people and

b wildlife?

5 Imagine that you and your family decide to visit the quarry for half a day. Describe what each member of your family might do during your visit.

Summary

The collection of natural resources can harm the landscape and affect people and wildlife. Attempts to improve damaged landscapes can be successful but may be expensive and take a long time.

17/2/09

What energy resources are there?

Resources can be defined as any material or product that we find useful. **Natural resources** are materials that we take from the environment. They include **physical resources** such as scenery, soils and water and raw materials that include minerals and fuel.

Resources can be divided into two groups.

1 **Non-renewable resources** are those which can only be used once and will eventually run out. Coal, for example, can only be burnt once.

2 **Renewable resources** can be used over and over again. Some, like the sun and wind, will never run out. Others, like soil and water, are renewable if they are not misused by people.

Energy resources

We all need energy to help us to work and to play. If we work or play hard we give off heat and use up energy. We have to eat and drink to replace this lost energy. Energy in nature also does work and can give off heat. Coal, for example, can be used to work machines and heat water.

Drawing **A** shows some non-renewable energy resources. So far, these have been quite easy and cheap to use. Coal, oil and natural gas are called **fossil fuels** because they come from the fossil remains of plants and animals. Unfortunately fossil fuels create a lot of pollution and are also causing changes in the world's climate.

A

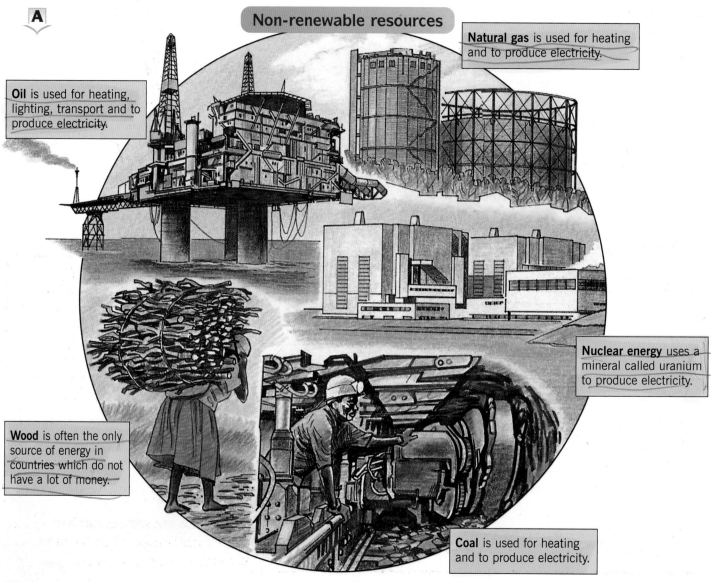

Non-renewable resources

Oil is used for heating, lighting, transport and to produce electricity.

Natural gas is used for heating and to produce electricity.

Nuclear energy uses a mineral called uranium to produce electricity.

Wood is often the only source of energy in countries which do not have a lot of money.

Coal is used for heating and to produce electricity.

Drawing **B** shows some renewable energy resources. These are mainly forces of nature, like water, wind and the sun, which can be used over and over again.

They tend to be difficult and expensive to use but cause little pollution and are a **sustainable** form of energy.

B

Renewable resources

Wind can turn windmills to create energy.

Wave energy is produced by wind blowing over the sea.

Solar energy comes from the sun.

Tidal energy can be produced where fast flowing tides enter river estuaries (mouths).

Water, if it is fast flowing and continuous, produces hydro-electricity.

Geothermal energy uses heat from inside the earth.

Activities

1 a What is a resource?

b Resources can be either natural or human. What are natural resources?

c Give three examples of natural resources.

2 Write down three differences between non-renewable and renewable energy resources.

3 a Wood is a non-renewable resource. Give reasons for this description.

b Wood could also be described as a renewable resource. Suggest how this might be.

4 Make a larger copy of table **C**. In each box put a tick ✔ for yes and an ✘ for no. The first one has been done for you.

C

	Used for heating	Used to generate electricity	A fossil fuel	Renewable resource	Non-renewable resource
Oil	✔				
Gas					
Coal					
Wood					
Wind					
Wave					
Sun					

Summary

Non-renewable resources can only be used once. They are usually cheaper to use but cause more pollution than renewable resources.

19/2/69

Electricity and the environment

We use electricity every day. It provides lighting and heating in our homes and offices. It powers computers, televisions and factory machinery. It is essential to modern-day living yet is usually taken for granted. All it needs is a touch of a switch. But do you know how electricity is generated and the harm it may cause?

Most of Britain's electricity is produced in thermal power stations. **Thermal power** is when electricity is produced by heat. Heat is usually obtained by burning fossil fuels such as coal, oil or natural gas.

Unfortunately, producing electricity in this way can cause problems for the environment. It can pollute our surroundings and use up resources that may never be replaced. Drawing **B** shows how electricity is produced in a thermal power station and the types of pollution it can create.

A A thermal power station at Drax, Yorkshire

B

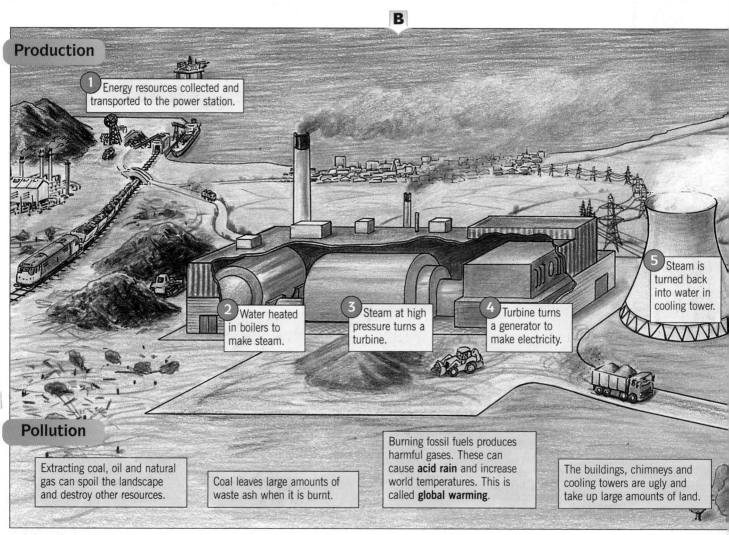

Production

1 Energy resources collected and transported to the power station.

2 Water heated in boilers to make steam.

3 Steam at high pressure turns a turbine.

4 Turbine turns a generator to make electricity.

5 Steam is turned back into water in cooling tower.

Pollution

Extracting coal, oil and natural gas can spoil the landscape and destroy other resources.

Coal leaves large amounts of waste ash when it is burnt.

Burning fossil fuels produces harmful gases. These can cause **acid rain** and increase world temperatures. This is called **global warming**.

The buildings, chimneys and cooling towers are ugly and take up large amounts of land.

I AM GOING TO SLEEP NOW — MORNING

How can pollution be reduced?

The electric companies are gradually replacing coal, the worst polluter, with oil and natural gas in their power stations. They have also developed ways of reducing emissions of harmful gases.

Another approach is to use renewable types of energy to produce electricity. As yet, these are neither as cheap nor as easy to use as fossil fuels. Although wind power is becoming increasingly popular, other forms of energy may have to wait until cheaper ways of using them are found or governments become more concerned about protecting the environment.

So several things can be done to reduce the harm done to the environment, but they all cost money. To pay for these things means putting up the price of electricity. Our problem is that we want cheap electricity and a clean environment. At present we have to choose between them.

C A wind farm at Rhyd-y-Groes, Anglesey

6 Power cables carry electricity to users.

Water is released from the cooling towers into the river. As this water is warm it can kill fish.

The pylons are unsightly and living too close to power lines may cause health problems.

Activities

1. Look at photo **A**. Make two lists of the effects that a coal-fired power station like Drax has on the environment:
   a local effects
   b wider effects.

2. Imagine that you live very close to Drax power station. What would you most dislike about living there? Give reasons for your answer.

3. Look at photo **C**. How would you feel about living close to that wind farm?

4. Look at the two opinions shown in cartoon **D**. Which do you think is more important? Give reasons for your answer.

D

It is better to have cheap electricity than to worry about the environment.

The protection of the environment is much more important than cheap electricity.

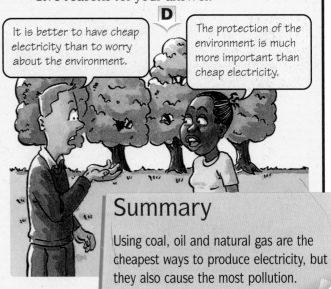

Summary

Using coal, oil and natural gas are the cheapest ways to produce electricity, but they also cause the most pollution.

Oil and the environment

The modern world has come to depend upon fossil fuels. Coal, oil and natural gas are relatively cheap sources of energy. Oil in particular is easy to obtain, cheap to transport and convenient to use. As we know, however, using fossil fuels can cause problems. Some ways that using oil can harm the environment are shown in **A** below.

Whilst the use of oil will always cause some problems, careful planning, management and the use of new technologies can help keep the damage to a minimum. Some methods that oil companies are using to protect the environment are shown in drawing **B**.

Study the disadvantages of oil and fossil fuels

A

1 Large areas of land have to be cleared when drilling for oil. This destroys vegetation and wildlife habitats and spoils attractive countryside.

2 Some oil is moved around by pipeline. Oil leaks caused by accident or terrorist attack can result in permanent damage to the environment.

4 Oil refineries convert crude oil to usable products like petrol, diesel and paraffin. These refineries are unsightly and cause air pollution.

5 Power stations burn oil to make electricity. Burning oil produces harmful gases that cause **acid rain** and contribute to **global warming**.

B

How can oil companies protect the environment?

◆ After use, restore damaged environments to their original state.
◆ Lay pipelines below ground and ensure regular safety checks.
◆ Choose pipeline and oil tanker routes carefully to avoid wildlife habitats and attractive scenery.
◆ Have emergency disaster plans in place to cope with oil spills.
◆ Paint refineries green or brown and screen behind trees.
◆ Develop cleaner fuels for industry and transport users.

3 Oil is also transported in large ships called oil tankers. Accidents to tankers cause huge oil spills which kill wildlife and pollute coastlines.

6 Oil that is turned into petrol and diesel is used to power cars, buses and lorries. The exhausts from these vehicles contain poisonous gases.

Activities

1 Look at the photos in **A**. Which examples:
  a show air pollution
  b damage only the immediate area
  c may have a worldwide effect
  d spoil the landscape
  e could be bad for your health?

2 Make a larger copy of the diagram below. Complete the diagram by adding examples to show how we damage the environment by using oil.

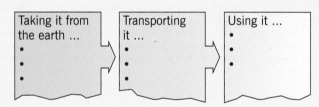

Taking it from the earth …	Transporting it …	Using it …
•	•	•
•	•	•
•	•	•

3 Choose any three of the photos in **A**. For each one:
  a Describe what you see.
  b Say what you think caused it.
  c Suggest how the problem might be reduced.
  Drawing **B** will help you.

Summary

Oil is one of the world's most important fossil fuels but its use can be damaging to the environment. Good planning and modern technology can help reduce some of the more harmful effects.

How can we conserve resources?

Many people are concerned about our environment. They worry that resources are being used up that cannot be replaced and our surroundings damaged and polluted by misuse. Most people agree that a way of solving the problem is through **sustainable development**. Sustainable development is a way of improving people's quality of life without wasting resources or harming the environment.

Sustainable development is largely the responsibility of world organisations and government bodies. There are, however, some things that can be done by individuals and local authorities. We can all, for example, develop a greater interest in our environment and take more care in how we all use its resources.

Some ways in which non-renewable resources can be conserved are shown in drawing **A**.

A

Reduce consumption
This means using fewer resources and being less wasteful, for example by switching off lights and computers when not in use and turning down central heating to save electricity.

Increase efficiency
This means using resources in the best possible way, for example by improving heat insulation in buildings, closing curtains at night and designing fuel-efficient planes, buses and cars.

Recycle
Recycling involves re-using materials rather than dumping them. Waste products such as paper, glass, metal cans, plastics and old clothes can all be processed and used again.

Activities

1 Look at drawing **A**. List the ways that **you** could help conserve resources. Try to give at least six.

2 Look at the headline below.
  a What do you think it means?
  b Why do you think it is a good way of dealing with the world's environmental problems?

Think global, act local

3 Make a larger copy of table **B**. Add examples from the classroom in drawing **C** for each method of resource conservation.

B

Reducing resource consumption	Increasing energy efficiency	Recycling

4 What could be done in your classroom to help improve resource conservation?

So are you doing your bit? Drawing **C** shows how sustainable methods can help conserve resources and save energy in the classroom.

Which of these methods apply to your classroom? What more could you do to save resources and help protect the environment?

C   Some ways of conserving resources in the classroom

Low-energy lights with automatic timers

Electrical equipment switched off when not in use

Door closed during lessons

Glass-fibre insulation in the ceiling

CONSERVING RESOURCES

Draught-proofing on door

Double-glazed windows

Waste material like paper, cans and plastics re-used or recycled

Wall cavities filled with foam

CLOTHES

Thermostats on radiators to control temperatures

Paper used sparingly

Carpet on floor

Books, equipment and furniture handled with care

Clothing recycled

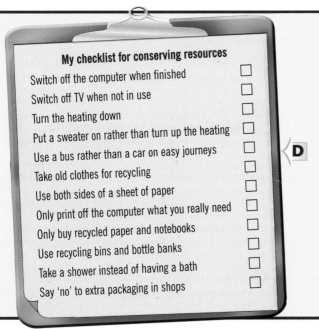

My checklist for conserving resources

Switch off the computer when finished ☐

Switch off TV when not in use ☐

Turn the heating down ☐

Put a sweater on rather than turn up the heating ☐

Use a bus rather than a car on easy journeys ☐

Take old clothes for recycling ☐

Use both sides of a sheet of paper ☐

Only print off the computer what you really need ☐

Only buy recycled paper and notebooks ☐

Use recycling bins and bottle banks ☐

Take a shower instead of having a bath ☐

Say 'no' to extra packaging in shops ☐

D

5 Make a copy of checklist **D**.

  a Tick the points in blue that you **already** do.

  b Tick the points in red that you **could** do.

  c Underline the points that you **will** do.

  d Describe how your decisions might help conserve resources.

Summary

Resources may be conserved by reducing consumption, increasing energy efficiency, and recycling. Sustainable development helps protect natural resources for the future.

Many environments need special protection and are looked after by government bodies or voluntary organisations. **The Wildlife Trusts** form the largest conservation group in the UK. The Trusts protect wildlife in towns and the countryside and rely upon donations of money to help do their job.

The Trusts help protect and conserve wildlife through planning and management. They:

◆ promote better access to wildlife areas

◆ work with schools and local communities

◆ provide information and organise events and clubs

◆ encourage a sustainable approach to wildlife conservation

◆ carry out improvements to nature reserves.

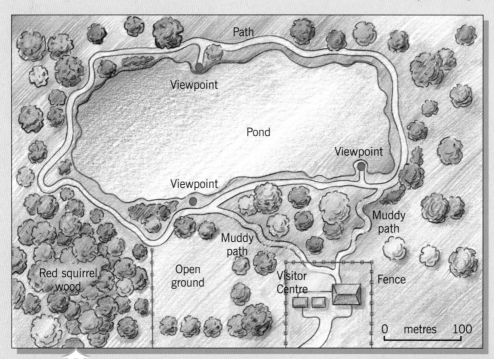

More information on the Wildlife Trusts can be found on their website at: **www.wildlifetrusts.org**

In this enquiry, you are a member of a team working for one of the Wildlife Trusts and have been given the task of improving a small nature reserve like the one in map **A** above. Your three main aims are:

◆ to improve facilities so that wildlife is protected and encouraged

◆ to improve public access within the reserve

◆ to improve visitor facilities and help promote enjoyment and understanding for the public.

The Heritage Lottery Fund has promised £325,000 towards the work, and a further £35,000 has been collected from Wildlife Trust members.

It will be best if you can work with a partner or in a small group. You will then be able to share views and discuss ideas with each other.

How can planning and management help protect the environment?

1 Look carefully at the suggestions on the next page and discuss how best to spend the money.

2 Draw up a table to show all the suggestions. Decide which you will use and how much will be spent on each one. Complete your table.

3 Write a report for the Project Director giving details of your plan. In your report describe what you propose to do and explain how each of your suggestions will help improve the reserve. Link your explanations to the three main aims.

4 Draw a poster encouraging people to visit your improved nature reserve.

B

£200 pays for 10 **red squirrel feeders**.

£200 pays for 20 **nest boxes** for fixing to trees.

Thorney Mere

£100 will print 5,000 **wildlife trail leaflets**.

£2 will plant **a tree or shrub**: 1,000 trees will turn an area 100 x 100 metres into woodland.

£100 will fund a **gate for disabled visitors**.

A medium-sized **hide for wildlife watching** costs about £5,000.

£5,000 pays for 100 metres of **boardwalk** in muddy, less accessible areas.

Should all paths be boardwalk, or just to the first viewpoint?

Do we save money for trees, squirrel feeders or nest boxes?

Do we put hides at all viewpoints?

How can we encourage visitors?

Improvements to the Visitor Centre will cost about £5,000 per room.

£50 pays for an after-school **environmental club meeting** or one **school visit**.

5 Population

People in the world

What is this unit about?

This unit is about people: where they live, how many there are and why they may move from one place to another.

In this unit you will learn about:

◆ **where people live in the UK**

◆ **what affects where we live**

◆ **world population distribution**

◆ **how world population changes**

◆ **the causes and effects of migration.**

Why is this population topic important?

Where we live and how many there are of us, affects us all in some way or another. Learning about population can help us understand some of the problems facing our world. It will also enable you to develop your own views on how those problems may best be solved.

Learning about population can help you:

◆ understand why some places are more crowded than others

◆ choose where you would like to live

◆ understand the reasons for population growth and movement

◆ appreciate the problems resulting from population growth and movement.

A | Bangladesh

B | Italy

C | London

◆ Of the places shown in photos **A**, **B** and **C**, which:
  – looks the busiest
  – looks the poorest
  – would you most like to live in
  – would you least like to live in
  – would you like to know more about?
  Give reasons for your answers.

◆ Describe the likely feelings of the migrants shown in photo **D**.

D

Are we evenly spread?

There are about 61 million people in the United Kingdom but where do they all live? Map **A** shows this. It is a **population distribution** map and shows how people are spread out across the country. You can easily see that the population is not evenly spread out. There are some areas with a lot of people and some with very few. The south and east seem to be most crowded, and the north and west the least crowded.

The map uses **density** to show how crowded places are. Density is the number of people in an area. It is worked out by dividing the total population by the total area and is usually given as the number of people per square kilometre. Places that are crowded are said to be **densely populated** and to have a high population density. Places with few people are said to be **sparsely populated** and to have a low population density.

The most crowded places of all are towns and cities. In Britain today almost 9 out of 10 people live in a town or city. Some towns and cities are shown on map **A** and in table **C**. London is by far the largest and most densely populated city in the United Kingdom. Almost 7 million people live there, and in the most crowded inner city areas there are up to 10,000 people in a square kilometre.

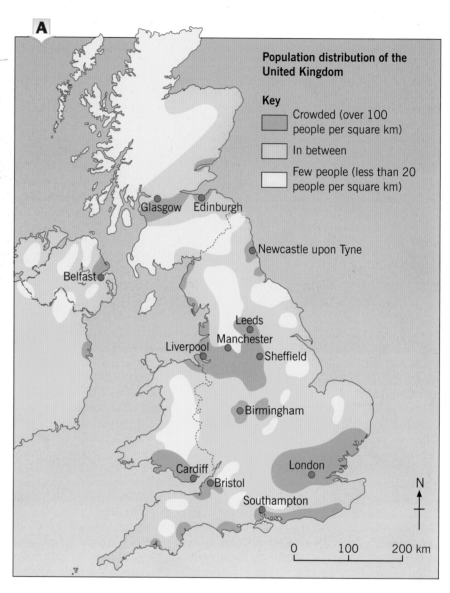

A Population distribution of the United Kingdom

Key

- Crowded (over 100 people per square km)
- In between
- Few people (less than 20 people per square km)

Glasgow · Edinburgh · Newcastle upon Tyne · Belfast · Leeds · Manchester · Liverpool · Sheffield · Birmingham · Cardiff · Bristol · London · Southampton

N

0   100   200 km

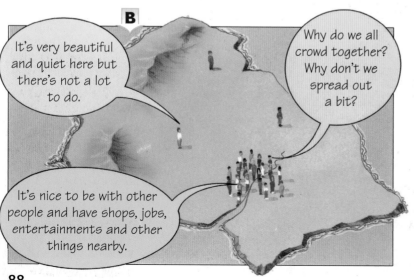

B

It's very beautiful and quiet here but there's not a lot to do.

Why do we all crowd together? Why don't we spread out a bit?

It's nice to be with other people and have shops, jobs, entertainments and other things nearby.

C

Population of some cities in Britain (figures are in thousands)

Belfast	280	Leeds	720
Birmingham	992	Liverpool	440
Bristol	380	London	7,513
Cardiff	310	Manchester	420
Edinburgh	450	Newcastle	259
Glasgow	560	Southampton	220

Note: Figures are for 2005 (estimated)

Photos **D** and **E** show places with very different population densities. Photo **D** is a typical city scene with many buildings, plenty of activity and a lot of people. Photo **E** was taken in Scotland. It shows part of the Highlands, a beautiful but sparsely populated area in the north.

Can you think why one place is crowded whilst the other has very few people? What is the area like where you live – is it crowded or is it sparsely populated? Can you suggest why it has that population density?

Activities

1 Copy and complete these sentences.

a A **population distribution** map shows ...

b **Population density** tells us ...

c **Densely populated** means that ...

d **Sparsely populated** means that ...

2 Map **F** shows the spread of population in Britain.

a Make a copy of the map and complete the key.

b Write a paragraph to describe the distribution of population. Include the following words in your description:

- spread   • unevenly   • south and east
- densely   • north and west   • sparsely

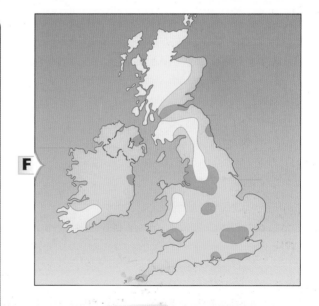

3 a List the cities from table **C** in order of size. Give the biggest first.

b Give three advantages of living in cities.

4 Look at sketch **B** and think carefully about what people need to live their everyday lives.

a Study photo **D** and make a list of the things that would help people to live there.

b Study photo **E** and suggest why very few people live in that area.

EXTRA

Draw a bar graph to show the number of people in each of the cities in table **C**.

- Arrange the bars in order of size with the biggest on the left.
- Use different colours for the cities in England, Scotland, Wales and Northern Ireland.
- Give your graph a title.

Summary

People are not spread evenly over Britain. Some areas are very crowded whilst others are almost empty. Population density is a measure of how crowded an area is.

What affects where we live?

Now where's the best place to live?

Not only is the distribution of population uneven in Britain, but it is uneven throughout the world. There are now over 6,000 million people in the world yet most of them live on only a third of the land surface. Like Britain, some areas are very crowded and others are almost empty.

There are many reasons for this. People do not like to live in places which are too wet or too dry, too hot or too cold. Nor do they like places that are mountainous, lack vegetation, are densely forested or liable to flood. People prefer pleasant places in which to live. They want to be able to earn money by working and have food available

through farming or from shops. They like to be near to other people and have things to do and places to go.

Factors that discourage people from settling in an area are called **negative factors**. Factors that encourage people to live in an area are called **positive factors**. Some of these are shown in the photos below and in diagram **G** on the next page.

Look carefully at the photos and for each one in turn try to work out why it is likely to be either a densely populated area or a sparsely populated area.

A Himalayan Mountains

B Amazon Forest

C Western Europe

D Sahara Desert

E Polar regions – Antarctica

F Bangladesh

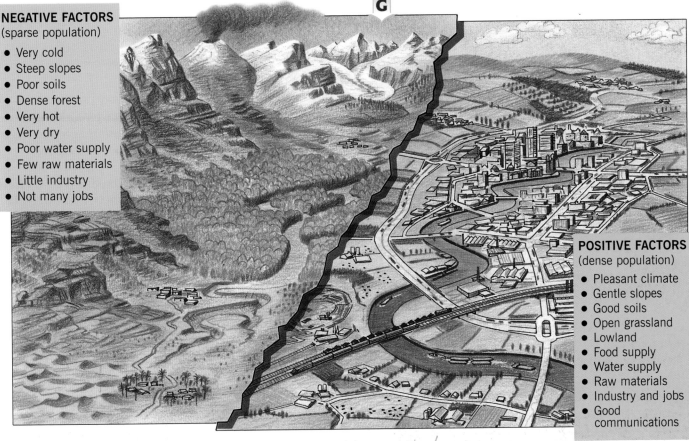

NEGATIVE FACTORS
(sparse population)

- Very cold
- Steep slopes
- Poor soils
- Dense forest
- Very hot
- Very dry
- Poor water supply
- Few raw materials
- Little industry
- Not many jobs

G

POSITIVE FACTORS
(dense population)

- Pleasant climate
- Gentle slopes
- Good soils
- Open grassland
- Lowland
- Food supply
- Water supply
- Raw materials
- Industry and jobs
- Good communications

Activities

H/w

1 a Which one of the photos **A** to **F** does this list of words and phrases best describe?

- steep slopes • snowy • very cold
- mountainous • icy • no soil
- no industry • very few people

b Which of these could be used to describe photo **F**?

- dry • steep • level • poor soils • wet
- good farming • factory work • cold
- hot • sparse population • many people

c Imagine that you are passing through the desert in photo **D**. Make a list of words and phrases to describe what it would be like. Try to give at least **eight** different things.

2 Copy table **H** and put the following into the correct columns:

- flat land • mountains • dense forest
- lowland • open grassland • good farming
- unreliable water supply • deep, rich soils
- thin, poor soils • job opportunities
- poor farming • deserts

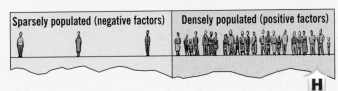

Sparsely populated (negative factors)	Densely populated (positive factors)

H

3 Give **two** reasons why few people live in

a mountain areas

b desert areas.

Summary

The way people are spread across the world is affected by many different things. These include relief, climate, vegetation, water supply, raw materials and employment opportunities.

EXTRA

Is the place where you live crowded or sparsely populated? What are the reasons for this? List the factors from diagram **G** which affect your area. Add any others that you think are important.

Where do we live?

The photo-map below is quite remarkable. It is made up from more than 37 million satellite images carefully put together to give a picture of the world. The red dots have been added to show the distribution of population. Look carefully and you can see many of the world's major features. The cold polar regions show up as white. The densely forested parts of South America and Africa are a lush green. The areas that are dry and lacking vegetation are shades of brown. Can you see the great mountain ranges? They show up as patches or streaks of white.

The map also confirms how unevenly people are spread over the world. Vast areas have hardly any people living in them whilst other areas seem to be very crowded. Try to name some of the emptiest places. Places with a lot of people include parts of Western Europe, India, China and Japan. Where else in the world does the photo-map show that there are a lot of people?

Photos of the six areas described below are shown on page 90.

A

Western Europe
Low-lying and gently sloping.
Pleasant climate.
Good water supply and soil for farming.
Easy communications and many resources for industry.
Densely populated.

Amazon Forest
Too hot and wet for people.
Dense forest makes communications and settlement difficult.
Sparsely populated.

Himalayan Mountains
Too cold for people.
Steep slopes are bad for communications and settlement.
Poor, thin soil unsuitable for crops.
Sparsely populated.

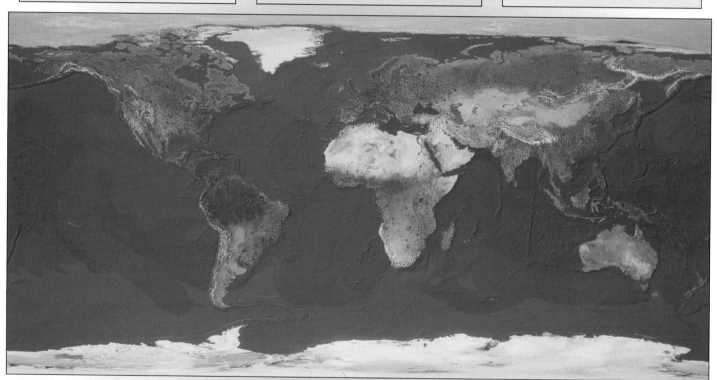

Polar regions – Antarctica
Too cold for people.
No soil for crops.
Snow and ice make communications and settlement very difficult.
Sparsely populated.

Sahara Desert
Too hot and dry for people.
Too dry and too little soil for crops to grow.
Sand makes communications difficult.
Sparsely populated.

Bangladesh
Low-lying and flat.
Rich, fertile soil. Hot and wet.
Ideal farming conditions.
Densely populated.

Cities are very popular places in which to live. They can provide housing, jobs, education, medical care and a better chance of getting on and enjoying life. More than half the world's population now live in cities and the number is increasing all the time.

The fastest growing cities tend to be in the poorer countries. Here, the urban population is expected to double in the next ten years. This will produce some very large cities.

One of these, Mexico City, is expected to become one of the three largest cities in the world by 2010. By then it will have a population of almost 30 million. At present its population is increasing by nearly half a million people a year. That is the same as all the inhabitants of Liverpool or Edinburgh suddenly arriving in Mexico City in a single year. Think of the problems that such a rapid increase must cause.

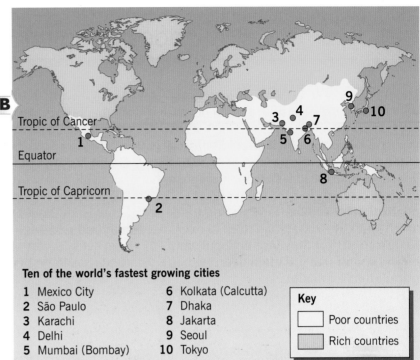

B

Tropic of Cancer

Equator

Tropic of Capricorn

Ten of the world's fastest growing cities

1 Mexico City
2 São Paulo
3 Karachi
4 Delhi
5 Mumbai (Bombay)
6 Kolkata (Calcutta)
7 Dhaka
8 Jakarta
9 Seoul
10 Tokyo

Key
☐ Poor countries
☐ Rich countries

Activities

H/W

1 Copy and complete the sentences below using the following words:

- densely
- deserts
- uneven
- polar regions
- sparsely

a The distribution of population over the world is _____

b The areas with the fewest people are the dense forests, _____ and _____ .

c Mountainous areas are _____ populated.

d Areas with good resources and industry are _____ populated.

2 Give named examples of:
a four densely populated areas
b six sparsely populated areas.

3 Which of the fastest growing cities:
a are in South America
b are in Asia
c is in a rich country?

4 Of the eight statements given below, three are correct. Write out the correct ones. The ten fastest growing cities are:
- mainly in poor countries.
- mainly in rich countries.
- mainly in polar regions.
- mainly between the tropics.
- on the coast.
- spread all over the world.
- in one continent.
- in South America and Asia.

5 Copy star diagram **C** and complete your diagram to show six reasons why people like to live in cities.

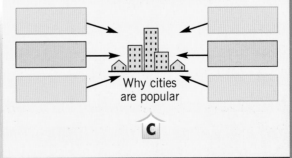

Why cities are popular

C

EXTRA

1 Use an atlas to name a country for each of the fastest growing cities in map **B**.

2 With help from an atlas, try to find out why central Australia is sparsely populated and east and south-west USA are densely populated.

Summary

People are not spread evenly over the world. Some of the most crowded places are in China, India, parts of Western Europe, and some areas of Africa and the USA. More and more people in the world are living in cities.

How does population change?

The population of the world is increasing very quickly. Experts have worked out that every hour there are an extra 8,000 people living on our planet. That is an increase of about 2 people every second or enough people to fill a city the size of Birmingham in about a week. In 1987 the world's population passed the 5,000 million mark and by the year 2000 it had risen to over 6,000 million. This increase in population is now so fast that it is often described as a **population explosion**.

A major world problem is how to feed, clothe, house, educate, provide jobs and care for this rapidly increasing population.

Graph **A** shows the changes in world population between the years 1100 and 2000. Notice how the population increase is not even. Until one hundred or two hundred years ago the population growth was actually very slow. Only in recent times has there been a real 'explosion'.

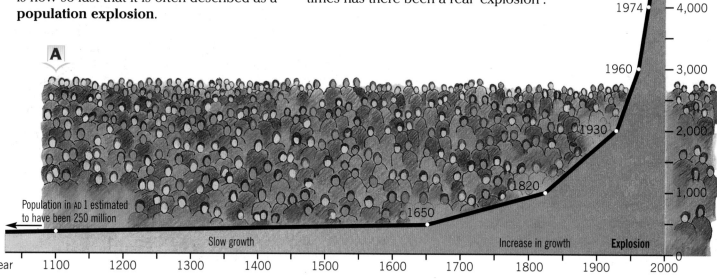

A

Population in AD 1 estimated to have been 250 million

Slow growth · Increase in growth · **Explosion**

Year 1100 1200 1300 1400 1500 1600 1700 1800 1900 2000

Millions of people: 6,000 / 5,000 / 4,000 / 3,000 / 2,000 / 1,000 / 0

Years marked: 2000, 1987, 1974, 1960, 1930, 1820, 1650

Population increases when the number of babies being born is greater than the number of people dying. The number of babies being born each year is called the **birth rate**. The number of people who die each year is called the **death rate**. Birth rates and death rates are measured as the number of births and deaths for each 1,000 of the population. The speed at which the population increases is called the **population growth rate**. Diagrams **B**, **C** and **D** show how the balance between births and deaths affects the population growth.

B

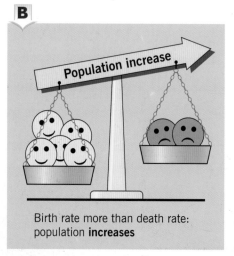

Birth rate more than death rate: population **increases**

C

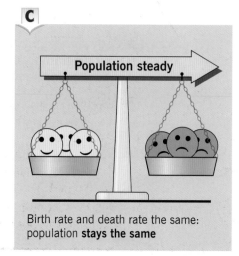

Birth rate and death rate the same: population **stays the same**

D

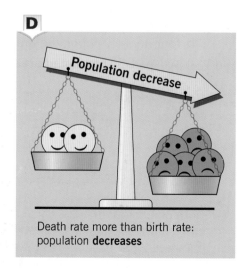

Death rate more than birth rate: population **decreases**

The population growth rate is not the same for all countries. In some, like Britain, the difference between birth and death rates is small so the population is changing only very slowly. In other countries, like Bangladesh, there are big differences between birth and death rates so the population is increasing rapidly.

Table **E** shows birth and death rates for some countries. Remember, the greater the difference between births and deaths, the larger the population change will be.

E

Country	Birth rate	Death rate	Natural increase
Bangladesh	30	9	21
Brazil	18	6	12
China	13	6	7
France	12	9	3
India	23	9	14
Italy	10	10	0
Japan	10	8	2
Mexico	22	5	17
UK	11	10	1
USA	14	8	6

- Figures given per 1,000 people. (2004)
- Poorer countries are shaded yellow.
- **Natural increase** is the difference between birth and death rates.

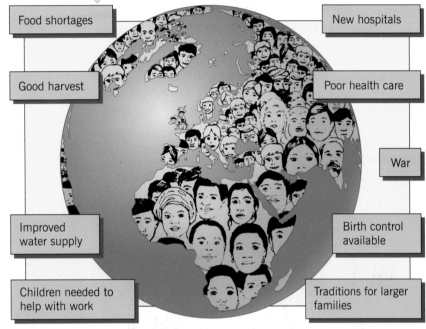

F **Things that can affect birth and death rates**

Food shortages

New hospitals

Good harvest

Poor health care

War

Improved water supply

Birth control available

Children needed to help with work

Traditions for larger families

Activities

H/W

1 a When did the world's population reach 1,000 million?

b How long did it take to double to 2,000 million?

c How long did it take to double again to 4,000 million?

2 Describe the increase in world population shown in graph **A**.

3 a Write a sentence to explain what each of the following terms means.
- Birth rate
- Death rate
- Population growth rate

b Why is 'explosion' a good description of population changes since 1950?

4 Copy and complete table **G** below by writing **increase**, **same** or **decrease** in the last column.

G

Births	Deaths	Population change
↗	→	
↗	→	
→	→	
↗	↘	

5 a List the countries from table **E** by the size of their natural increase. Put the one with the greatest increase first.

b What do you notice about the richer and the poorer countries?

EXTRA

1 Draw table **H** below and from diagram **F** sort the things that affect birth rates and death rates into the correct columns.

H

Birth rate		Death rate	
High	Low	High	Low

2 Suggest **three** reasons why people in the UK may be more likely to live longer than people in poorer countries.

Summary

The world's population is increasing at a very rapid rate. Growth is very much faster in the poorer countries than in the richer ones. Population change in a country depends mainly on the birth and death rates.

What is migration?

London is the UK's largest city with a population of just over 7.5 million. After a period of decline, London's population is slowly growing again. The growth is expected to continue for many years. This is partly because of natural increase but mainly because, like most cities, it acts like a magnet and attracts people from other places to live there.

People who move from one place to another to live are called **migrants**. They have a big effect on populations because they increase the numbers and can alter the mix of people who are living in a place.

Graph **A** shows London's growth since 1961. The photos in **C** show some of the people who have moved into London. Notice how different the people look and what varied backgrounds they have. Their comments help to explain some of the reasons why people migrate.

A

Population of London

Millions of people

Estimated

8.5		
8.0		
7.5		
7.0		
6.5		
1961	1981	2001 2021

Year

B London

C

Hamish

I'm from a small village in Scotland and suffer from poor health. My family live in London so I've come to join them. It will also be easier to get medical care here.

Caroline

I went to university in Bristol but I'm ambitious and think that I'll have a better chance of getting a good job and enjoying life in London.

Janine

I qualified as a nurse in New York. I worked in Mexico for a while but have now moved to London to gain more experience and make new friends.

Chang

I came here from China. My company in Shanghai sent me to London some years ago and I like it so much that I'm going to stay.

Carmel

I've lived here all my life but my family are originally from Jamaica. They came here in the 1950s when the British government asked for people to fill job vacancies.

Zaric

I was brought up in a small town in Croatia. My family were killed in the war there and I always felt in danger. I moved to London to start my life again.

Migration is when people move home. The movement may be just around the corner to a better house. It may be from one part of the country to another in search of a job. It might be from one country to another for a different way of life. For many people country areas have very little to offer so they move to the towns and cities. This is called **rural-to-urban migration**. For other people a move to a different country holds many attractions. This is called **international migration**.

People migrate for two reasons. Firstly, they may wish to get away from things that they do not like. These are called **push factors** and include a shortage of jobs and poor living conditions. Secondly, people are attracted to things that they do like. These are called **pull factors** and include pleasant surroundings and good medical care.

PUSH FACTORS
- Political fears
- Not enough jobs
- Few opportunities
- Natural disasters
- Wars
- Unhappy life
- Shortage of food

PULL FACTORS
Hope for
- Better way of life
- Chances of a job
- Improved living conditions
- Education
- Better housing
- Medical care
- Family links

D

Activities

1 What do each of these terms mean?
- Migration
- Rural-to-urban migration
- International migration

2 **a** Copy and complete bar graph **E** to show population growth in London.

b Describe the change in population since 1960.
- What is the overall change?
- Is the recent change slow, steady or rapid?

c Suggest reasons for the change.

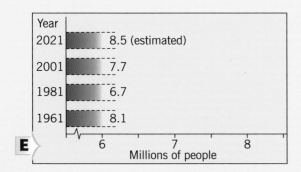

Year		
2021		8.5 (estimated)
2001		7.7
1981		6.7
1961		8.1

Millions of people
E

3 **a** What is a push factor? Give two examples.

b What is a pull factor? Give two examples.

4 Copy table **F** and use it to explain why the people in **C** migrated to London.

F

	Push factor	Pull factor
Hamish		
Janine		
Carmel		
Caroline		
Chang		
Zaric		

5 Imagine you are a migrant like Zaric in **C**.

a You have just arrived in London. Describe your first few days there. Think about language difficulties, getting a job, food, shelter, making friends ...

b Describe why you migrated from Croatia to London. Include push and pull factors.

Summary

Migration is the movement of people from one place to another. This movement may be the result of push and pull factors. The migration of people affects population size and the variety of different people in a place.

97

Who migrates to the UK?

Look at the welcome signs on poster **A**. They all mean the same but are in different languages. The poster is from Newham in London where many of the residents are newly arrived and have a language other than English.

In fact people have been arriving in Britain from other countries for more than 2,000 years. Some have come as invaders, some to escape problems in their own countries and some simply to find jobs and enjoy a better way of life.

Indeed the UK population is made up of **immigrants** and has always been a country of mixed races and cultures. The majority of UK residents are descended from Romans, Vikings, Angles, Saxons and Normans. The Irish have also settled in Britain for several centuries, while many other Europeans migrated here both during and after the Second World War.

Some of the largest groups of more recent immigrants are from countries that were once part of the British Empire, like India, Pakistan, Bangladesh and the West Indies. After the war there were serious labour shortages in Britain. The government of the time invited people from these countries to come and fill job vacancies. Almost 1 million responded to the call. Most immigrants settled permanently in Britain. They had families and, along with their descendants, became UK citizens.

The graphs below show where people now living in the UK were born. More than 1 in 12 (4.9 million) were born overseas.

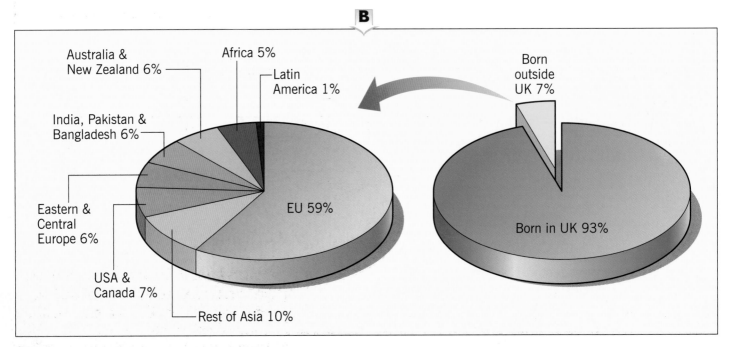

In recent years there has been an increase in the number of migrants who are **asylum seekers**. These are people who live in danger in their own country and want to move to a place where they will be safe. Most asylum seekers are homeless when they arrive in the UK. The government houses them in temporary accommodation until it decides whether they will be allowed to stay.

Some people are against asylum seekers. They say that they are illegal immigrants and will ruin the country. Others are in favour of them. They argue that most are genuine people with a need for asylum and a right of access to the UK. They point out that most are well educated and are able to make a positive contribution to UK life.

There is concern, however, that there are just too many migrants coming into the country. Whilst these people generally bring benefits, if there are too many of them, there can be problems. For this reason, the government has begun to restrict and control the number of immigrants entering Britain. Since 2003 only those with specific jobs to go to, with descendants already in the country, or genuine **refugees**, are now allowed access.

C

Asylum seekers arriving in the UK

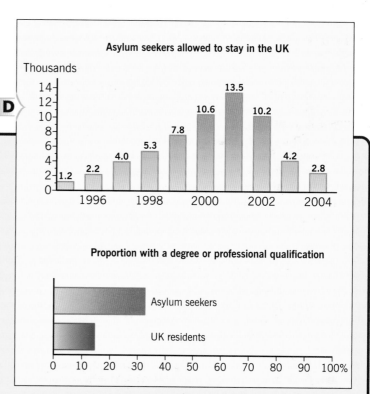

D

Asylum seekers allowed to stay in the UK

Thousands

14
12
10
8
6
4
2
0

1.2 2.2 4.0 5.3 7.8 10.6 13.5 10.2 4.2 2.8

1996   1998   2000   2002   2004

Proportion with a degree or professional qualification

Asylum seekers

UK residents

0 10 20 30 40 50 60 70 80 90 100%

Activities

1 a How many different languages are shown on poster **A**?

   b Write out the welcome signs from three European countries and three non-European countries.

2 Make a list of where people in your class come from. Try to go back to previous generations. Sort the list under the headings on graph **B**.

3 Look at graph **B**.

   a Where have most immigrants come from?

   b Suggest why immigrants are allowed into the UK from this area.

   c Which group of immigrants are likely to have relatives in the UK already?

4 Write a short letter to a newspaper supporting asylum seekers. Zaric on page 96 and the graphs **D** will help you.

Summary

The UK is made up of people from many different countries. In recent years, most migrants have come either from the EU or from countries that were once part of the British Empire.

What are the effects of migration?

The effect of migration into Britain has been considerable. It has caused an increase in numbers and altered the mix of people in the country. It has also produced a **multicultural society** where people with different beliefs and traditions live and work together. Most people agree that this has added variety and interest to the UK.

There are many other effects of migration. Some affect the migrant and others affect the place they have moved from as well as the place they have moved to. Look at the comments in **A** and **B** which show some of these effects.

A Migration – the good ... and the bad

The Singh family
It's great here... we love it! Sanjit got a job straight away in a clothing factory in Newham. We have a small flat and the children are at a local school. We are all making progress with the language and have already made lots of friends. Migration was a good move for us ... we are really very happy.

The Khafoor family
I wish we had never left Pakistan. We can't speak the language very well and have had great difficulty getting work and making friends. It's been hard finding a reasonable place to live because we have so little money and people just don't seem to want to help us. Migration has made us homesick and very unhappy.

Paul Richards, restaurant owner
I think the immigrants are great. I can't afford to pay high wages but the immigrants seem to be happy on low pay, so that's good for my business and I'm happy to employ them. I think we should encourage immigrants to live in the UK because they do jobs that we don't like and they are good for the economy.

Dave Parker, unemployed
I'm sick of immigrants coming into our country. They take our jobs, live in our houses and try to change our cities to suit their way of life. I don't have a job and I really don't think it's fair that a foreigner should get work before me.

B

Activities

1 a Give five reasons why the Singh family are pleased they migrated to the UK.

b Give five reasons why the Khafoor family are unhappy about migrating to the UK.

2 Look at the list of effects of migration in **C** below. Arrange the statements in a diamond shape as shown in drawing **D**.
- Put the advantage you think is the most important at the top.
- Put the next two below, and so on.
- The greatest disadvantage will be at the bottom.

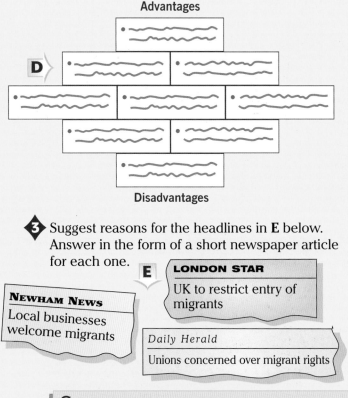

Advantages

D

Disadvantages

3 Suggest reasons for the headlines in **E** below. Answer in the form of a short newspaper article for each one.

E **LONDON STAR**
UK to restrict entry of migrants

NEWHAM NEWS
Local businesses welcome migrants

Daily Herald
Unions concerned over migrant rights

C
- Puts pressure on health services and schools
- Provides opportunities for migrant families
- Increases the country's wealth
- Causes religious and cultural problems
- Takes jobs from UK workers
- Adds interest and variety
- Provides needed skilled workers
- Causes overcrowding in some cities
- Helps people understand other ways of life

Summary

Many immigrants settle happily in their new surroundings but for some there can be difficulties. Migrants can be a great help to the economy. They provide much needed skills and add variety and interest to the UK.

How can we compare local areas?

This unit is about people: where they live, how many there are and how some of them move from one place to another. So what about the people in the area where you live? How many are there, how crowded is it and where do they come from? Learning about other places can also be interesting. What are they like and how do they compare with the place where you live?

Information about your local area can easily be found on the internet. The government's **National Statistics** website gives facts and figures for the whole country. Typing in your postcode or town name will give you information about where you live. You can easily find out about other areas using the same method. You will also be given UK averages to compare with your own.

The website has statistics on a variety of population topics including population change, population density, age groups and ethnic origin.

It even lists the most popular names for babies! Much of the data is presented as graphs to make it clearer and help you make comparisons.

The information on these two pages is for the London boroughs of Newham and Chelsea. Notice how different they are. Which one is most like your local area? Which one would you prefer to live in?

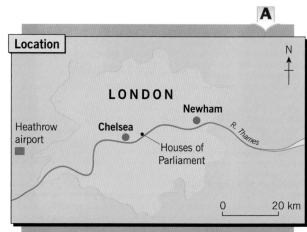

A

Location

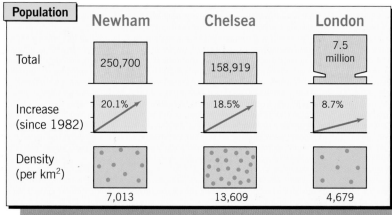

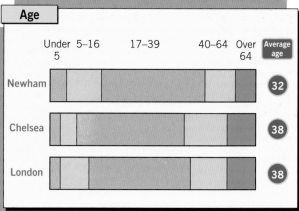

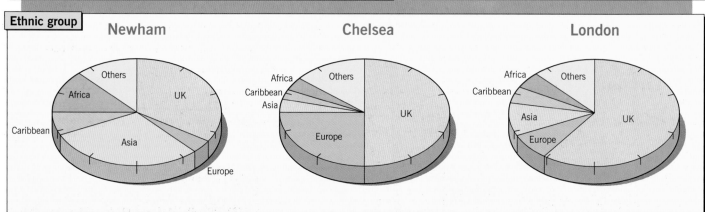

B

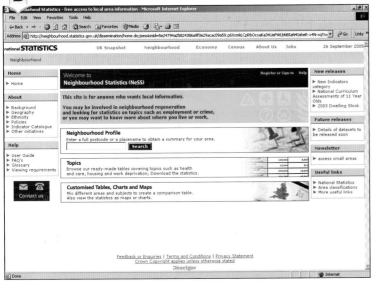

C Street market in Newham

D

	Newham	Chelsea
Closest to the city centre		
Has most people		
Has faster growth than London		
Is most crowded		
Has most young people		
Has most at retirement age		
Has more over 40s than London		
Has fewest from Europe		
Has greatest mix of nationalities		
Likely to attract more migrants		

Activities

1 a Make a larger copy of table **D**.

 b Tick the correct column for each statement. More than one column may be ticked.

2 a Copy table **E** below and complete it using information from the 'Ethnic group' graph in **A**. The first item has been done for you.

E

Ethnic groups		
	Newham	**London**
Highest (%)	1 UK (34%)	1
	2	2
	3	3
	4	4
	5	5
Lowest (%)	6	6

 b Look at your completed table and describe the main differences between Newham and London as a whole.

3 You work for the council and have been asked to produce a report comparing your local area with the UK. Your report is about the population and people who live in the area.

 a Go to the Nelson Thornes website at: www.nelsonthornes.co.uk/keygeography.

 b Log on to the **National Statistics** website.

 c Click '**Neighbourhood**' at the top of the screen. You should have a page like the one in **B**.

 d Enter your postcode or town name.

 e Click '**People and society**'.

 f Now collect information for your report.

Your report should present some information as graphs, maps and tables. Remember to explain what it shows and compare it with the UK averages.

Summary

Information about population and people may be found on the internet. This can be used to compare your local area with other places.

The population enquiry

As we have seen earlier in this unit, people are not spread evenly around the world. Some places are crowded whilst other places have very few people. There are many reasons for this. Some are **positive factors** which encourage people to settle in an area. Others are **negative factors** which discourage settlement. There are almost always some positive factors and some negative factors working in an area. It is the balance between the two that determines how crowded a place becomes.

Look at drawing **D** on the next page which shows an imaginary continent in the northern hemisphere. It shows many of the factors affecting population distribution that were described on pages 88 to 93. Your task in this enquiry is to determine the most likely population distribution on the continent using negative and positive factors.

How can positive and negative factors affect where people live?

1 a Make a larger copy of table **B**.

b List the positive and negative factors for each of the areas labelled 1 to 10 on drawing **D**.

c For each place, decide if it is likely to be crowded, to have few people, or to be in between. Write your decision in the 'Population density' column.

d In the last column of the table, give reasons for your suggested population density.

2 a Make a larger copy of map **C**.

b Show the population distribution of the continent by dividing it into areas that are likely to be crowded, have few people or be in between. Use colours similar to those on map **A** on page 88.

c Choose the most likely locations of three main towns and mark them on your map.

d Complete your map by colouring the coastline blue and giving names to the towns, mountain ranges, rivers, seas and imaginary continent.

3 Describe and suggest reasons for the locations that you have chosen for the three main towns.

Area	Positive factors	Negative factors	Population density	Reasons
1				
2				
3				
4				

B

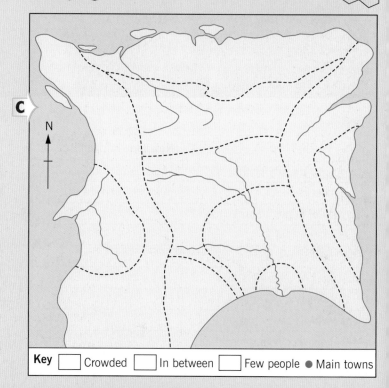

C

N

Key ☐ Crowded ☐ In between ☐ Few people ● Main towns

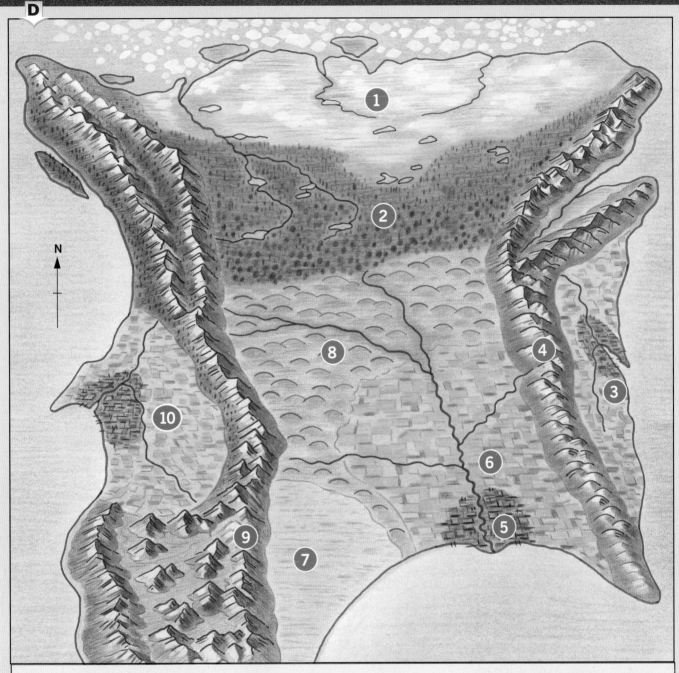

D

N

1. Mainly flat and low-lying land. Poor, thin soils. Little vegetation. Very cold.

2. Hilly with many lakes. Poor soils. Dense coniferous forest. Very cold winters.

3. Flat or gently sloping land. Raw materials available for industry. Fertile soils. Warm climate with plenty of rain.

4. Steep slopes. Thin soils. Few jobs. Poor communications.

5. Flat and low-lying land. Serves rich farming area inland. Industries and port facilities.

6. Flat or gently sloping land. Good-quality farmland. Warm summers with plenty of rain.

7. Very hot and dry. Mainly desert with few rivers. Little agriculture.

8. Mainly hilly with poor soils. Large-scale cattle farming. Few jobs. Cold winters.

9. Rugged, mountainous area. Difficult communications. Few jobs. Poor, thin soils. Wet.

10. Flat and low-lying. Fertile soils and good farming. Industries and port facilities. Good climate.

Kenya, a developing country

What is Kenya like?

What is this unit about?

This unit is about Kenya, a developing country. Although it is very poor economically, it is rich in scenery and wildlife and has cheerful people and some great athletes.

In this unit you will learn about:

◆ Kenya's main physical features
◆ population distribution and movement
◆ differences in urban and rural life
◆ sustainable development in Kenya
◆ the features of a developing country.

Why is learning about Kenya important?

Learning about Kenya gives you an interest and knowledge of people and places that are very different from those found in the UK.

This unit will help you to:

◆ broaden your knowledge of our world
◆ learn about different landscapes and climate
◆ understand ways of life that are different from your own
◆ understand differences in development.

A | Mt Kilimanjaro from Amboseli National Park

B Nairobi

C Samburu village

◆ Compared with the area where you live, how different is:
  – the countryside in photo **A**
  – the town in photo **B**
  – the village in photo **C**?

◆ Make a list of at least six words to describe the Kenyans shown in photo **D**.

D Kenyan schoolchildren

What are Kenya's main features?

'Jambo' means hello and is the greeting always given by Kenyans. It is usually accompanied with a broad smile and an offer of help. Indeed most Kenyans seem always to be cheerful, relaxed and willing to help others.

Yet life in Kenya can be very difficult. It is one of the poorest countries in the world, has few services, little industry and a low **standard of living**. It is called a **developing country** and is very different from the UK which is rich, has a high standard of living and is an example of a **developed country**.

Despite being poor in terms of wealth, Kenya is rich in scenery and wildlife and has become a popular tourist destination. Inland there are high mountains, volcanoes, lakes and grassy plains. On the coast there are long sandy beaches, coral reefs and tropical forest.

Most tourists are attracted by the wildlife. They go on organised tours called **safaris** and view the animals from open-topped mini-buses. After a safari, many visitors travel to the coast for a more relaxing beach holiday. Because Kenya lies on the Equator, the climate is usually hot and sunny throughout the year.

Kenya also has a varied population which is made up of people from several different groups or tribes. Each group has its own **ethnic** background with its own language, religion and way of life. Two of the best known tribes are the Maasai and the Kikuyu.

Kenya's two major cities are Nairobi, the capital, and Mombasa on the coast.

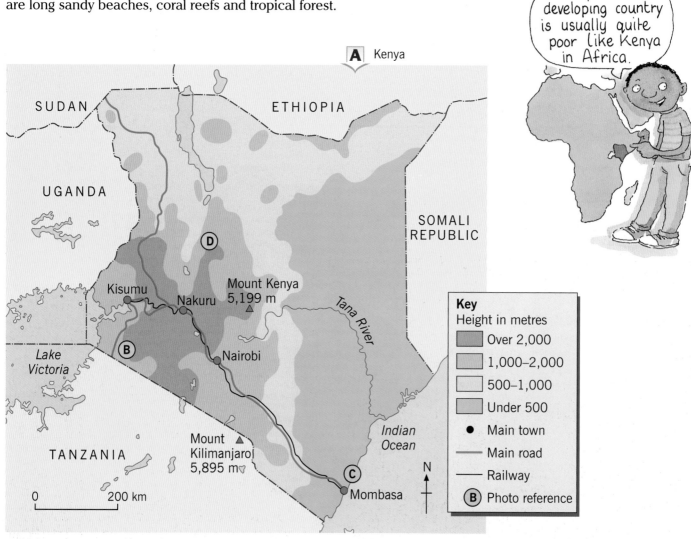

Jambo! A developing country is usually quite poor like Kenya in Africa.

A Kenya

Activities

1 Give the meaning of the following terms. The Glossary at the back of the book may help you.

a developing country   **b** developed country

c standard of living   **d** ethnic

2 Look at map **A** opposite.

a Name the five countries that border Kenya.

b Name the four main towns in order of height above sea level. Give the highest first.

c How long is Kenya's coastline?

d How far is it by rail from Kisumu to Mombasa?

e What is the furthest distance from north to south?

3 Copy and complete quizword **E**, using the following clues.

ⓐ Kenya's highest mountain

ⓑ A line of latitude across Kenya

ⓒ An ocean off Kenya's east coast

ⓓ A holiday where wild animals are viewed

ⓔ Kenya's main port

ⓕ A traditional Kenyan greeting

ⓖ A Kenyan tribe

Make up a clue for downword ⓗ.

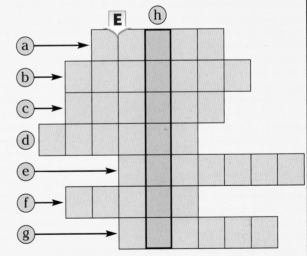

4 Imagine you are on holiday in Kenya. Use the information on this page to help you write a postcard home which describes the country.

Summary

Kenya is a developing country. Although it is not rich, it does have a wealth of spectacular scenery and wildlife.

What are Kenya's main physical features?

Most visitors to Kenya fly in to Nairobi, the country's largest city. Nairobi is in a region called the Central Highlands which millions of years ago was an area of considerable **volcanic activity**.

During that time huge cracks developed in the earth's surface. Two of the cracks ran from north to south through Kenya and as the land between the cracks collapsed, a huge **rift valley** was formed. Earth movements and volcanic eruptions are now quite rare but the cone-shaped remains of volcanoes can be seen throughout the area. Mount Kenya at 5,899 metres is the highest and best known of these.

The scenery and vegetation change dramatically towards the coast. East of the Central Highlands the land is more gently sloping and there are huge areas of grassland. A band of lush, green, tropical **rainforest** lies along the coast where the climate is very hot and wet.

A Mount Kenya

B

Mount Kenya is an old volcano. Despite being on the Equator, it is high enough to be snow-covered all year. It even has its own glacier.

The north and east of Kenya is much flatter than the centre. There is little rainfall here, so much of the land is desert with little vegetation.

The Great Rift Valley has steep sides, is very deep and in places is over 60 km wide. The valley stretches almost the whole length of Africa.

Cracks in the earth's crust

Grassland covers much of the coastal plain. The National Parks in this area are rich with wildlife.

The east coast has long sandy beaches. Coral reefs lie offshore and protect the coast. Coral thrives in the warm, clear waters of the Indian Ocean.

Lake Turkana

Lake Victoria

Mount Kenya

Nairobi

Equator

SOMALI REP.

River Tana

Indian Ocean

Mombasa

N

Kenya is on the Equator which means that average temperatures are high throughout the year. There are no winters or summers as there are in Britain. In terms of temperature, one day is very similar to the next. Seasons are defined by rainfall amounts. The rainy season is usually from April to May.

A typical day in Nairobi begins with bright, clear weather. By the afternoon clouds will have built up and there may be a shower of rain. In the evening it becomes clear and dry again, though rather cool.

Not all of Kenya is like this. The north is hot and dry and rarely has rain. Further south and along the coast there is plenty of sunshine, high temperatures throughout the year, and more rainfall. Afternoon sea-breezes cool the air on the coast but evenings are warm and very pleasant.

With attractive scenery and a climate like this, it is not surprising that Kenya's coastline has become increasingly popular with tourists.

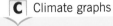 **C** Climate graphs

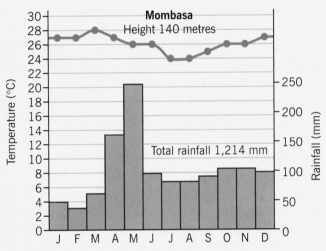

Activities

1 a What is the Great Rift Valley?

b With the help of the diagrams below, explain how the rift valley was formed.

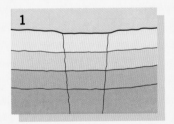

2 Imagine that you have just taken photo **A** of Mount Kenya. Write a caption for the photo to both describe **and** explain the mountain's main features. Include the following:

- steep
- rocky
- volcano
- Equator
- cold
- cone-shaped
- snow-covered
- 5,899 m high

3 Look at the climate graph for Nairobi above.

a Which three months are the wettest?

b How much rainfall is there in July?

c What temperature would you expect in July?

d Why do you think Nairobi is cooler than Mombasa?

4 a Describe the attractions for tourists to Kenya's coastline. Use the following headings:
- General weather
- Coastal landforms
- July temperature
- Vegetation cover
- July rainfall
- Sea conditions.

b When is a bad time to take a holiday in the area? Give reasons for your answer.

Summary

Many of Kenya's most attractive landforms can be seen either in the Central Highlands or along the coast. Most of Kenya is hot all year but a lack of rain can be a problem.

Why is Kenya's population unevenly spread?

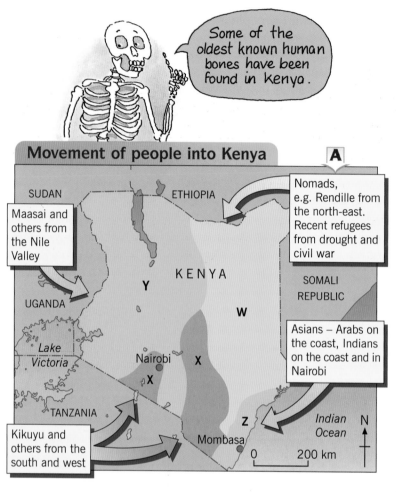

Some of the oldest known human bones have been found in Kenya.

Movement of people into Kenya **A**

SUDAN

ETHIOPIA

Maasai and others from the Nile Valley

Nomads, e.g. Rendille from the north-east. Recent refugees from drought and civil war

K E N Y A

Y

UGANDA

W

SOMALI REPUBLIC

Lake Victoria

Nairobi

X

X

Asians – Arabs on the coast, Indians on the coast and in Nairobi

TANZANIA

Z

Indian Ocean

N

Kikuyu and others from the south and west

Mombasa

0   200 km

The population of Kenya, as in most other countries, is not spread out evenly. Some places are very crowded while others have very few people living there. This is mainly due to:

◆ **migration** – where the people of Kenya originally came from

◆ **physical conditions** – differences in climate and relief in Kenya.

The movement of people into Kenya

Most present-day Kenyans are descended from African tribes who arrived in the country from four main directions (map **A**). Most came from the south, making this the most densely populated area.

While individual tribes still remain today, their way of life has been changed through marriages and contact with people from other ethnic groups.

Differences in physical conditions

Although Kenya is not a very large country, the physical conditions vary considerably from place to place. Map **B** shows how the distribution of population is affected by differences in rainfall, temperature, water supply, relief, soil and vegetation.

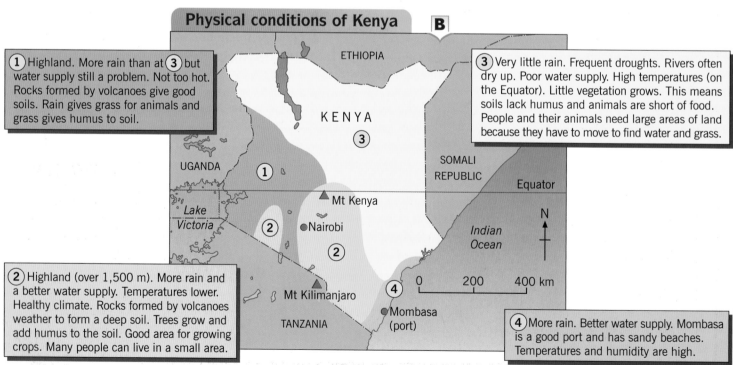

Physical conditions of Kenya **B**

ETHIOPIA

① Highland. More rain than at ③ but water supply still a problem. Not too hot. Rocks formed by volcanoes give good soils. Rain gives grass for animals and grass gives humus to soil.

③ Very little rain. Frequent droughts. Rivers often dry up. Poor water supply. High temperatures (on the Equator). Little vegetation grows. This means soils lack humus and animals are short of food. People and their animals need large areas of land because they have to move to find water and grass.

K E N Y A

③

UGANDA

①

SOMALI REPUBLIC

Equator

Lake Victoria

▲ Mt Kenya

②

● Nairobi

②

Indian Ocean

N

② Highland (over 1,500 m). More rain and a better water supply. Temperatures lower. Healthy climate. Rocks formed by volcanoes weather to form a deep soil. Trees grow and add humus to the soil. Good area for growing crops. Many people can live in a small area.

▲ Mt Kilimanjaro

④

0   200   400 km

TANZANIA

● Mombasa (port)

④ More rain. Better water supply. Mombasa is a good port and has sandy beaches. Temperatures and humidity are high.

Present-day movements of population

In all developing countries there is a large movement of people from the countryside to the towns. This is called rural-to-urban migration. In Kenya it is mainly the Kikuyu who move. Their traditional home is in area ② on map **B**.

When driving through rural Kikuyu countryside it is hard to see why they want to move. It is one of the few parts of Kenya with roads, it has the best farmland and water supply in the country and the environment appears clean and pleasant. To those living there, especially those at school or just starting a family, it is less attractive.

Diagram **C** gives some of the reasons why many Kikuyu want to move to Nairobi, the capital of Kenya.

C

Nairobi has big, modern buildings which include hospitals, shops, cinemas and a university.

Kenya has one of the highest birth rates in the world. The average family size is 7.6 people. There are too many of us to find jobs on the farms and in the shambas.

The Kikuyu have lived in villages and small towns for a long time. The change to city life should be easy.

We are farmers but many of us do not own any land and if we do, the plots are very small.

We have learned some skills at school but we cannot use them in our local village.

It is not far to Nairobi so we can get work there and still visit our village.

Activities

1 Read the following statements about Kenya's population. Write out the four statements which are correct.
- Kenya's population is spread evenly.
- Kenya's population is not spread evenly.
- The Maasai came from the Nile Valley and live in the south-west.
- The Maasai came from the south-west and live in the Nile Valley.
- The Kikuyu live in the north.
- The Kikuyu came from the south and west and live on higher land.
- Arabs and Indians live on the coast near Mombasa.

2 Make a copy of diagram **D** and complete it to show why the south of Kenya is more crowded than the north. For each box choose the correct word from the two given in brackets. 'Relief' has been done for you.

3 Imagine that you live in a small Kikuyu village and are about to migrate to Nairobi.

a Give at least three reasons for leaving your village.

b Give at least three advantages of living in Nairobi.

Diagram **C** will help you to answer this question.

D

Relief	Rainfall	Water supply			Temperature	Soils
Low			NORTH	Few people		
High			SOUTH	KENYA Many people		
(High/Low)	(High/Low)	(Good/Poor)			(Hot/Warm)	(Good/Poor)

Summary

The distribution of population in Kenya is mainly affected by physical factors such as climate, water supply, relief and soil. Most people live in or near to the capital city of Nairobi.

What does Nairobi look like to newcomers?

Like most cities in developing countries there are two sides to Nairobi. One side is seen by overseas visitors and the relatively few wealthy Kenyans. Photo **A** shows the most important building in Nairobi. It is the Kenyatta International Conference Centre, named after Jomo Kenyatta, the first president of Kenya. Around it, in central Nairobi, are tall, modern buildings and wide, tree-lined streets.

The other side of Nairobi is the one seen by most of the Kenyans who migrate from the surrounding rural areas. Many migrants move to live with family and friends already living in Nairobi. They share houses, food and even jobs. In time the newcomers may be able to construct their own homes in one of several **shanty settlements** found on the edge of the city (photo **B**).

A

B

C

Living in Kibera

Kibera is one shanty settlement. It is 6 kilometres and a 7 pence bus ride from the city centre. However, most inhabitants who wish to make that journey have to walk because they cannot afford the fare.

Houses in Kibera are built close together. Sometimes it is hard to squeeze between them. The walls are usually made from mud and the roofs from corrugated iron (photos **C** and **E**). Inside there is often only one room. Very few homes have water, electricity or sewage. One resident, who is considered rich by Kibera standards, has a tap and can sell water to his neighbours. He also has a toilet but this is only emptied four times a year.

Sewage runs down the tracks between the houses (photo **C**). In the wet season rain mixes with the sewage making the tracks unusable, so small children are kept indoors for several weeks. In the dry season the tracks become very dusty. Photo **D** shows an open drain along a main sidetrack being cleared. What might happen to the sewage the next time it rains?

Kenya has one of the highest birth rates in the world. In Kibera it is not unusual for families to have more than ten children. Very few children can read or write as there is only one small school. Many suffer from a poor diet or malaria. Others catch diseases by drinking dirty water or playing in sewage (photo **C**).

People have to find their own way to earn money. Some have small stalls from which they sell food. Others, who have learnt a skill, may turn their houses into shops (photos **B** and **E**) or collect waste material and recycle it in small workshops.

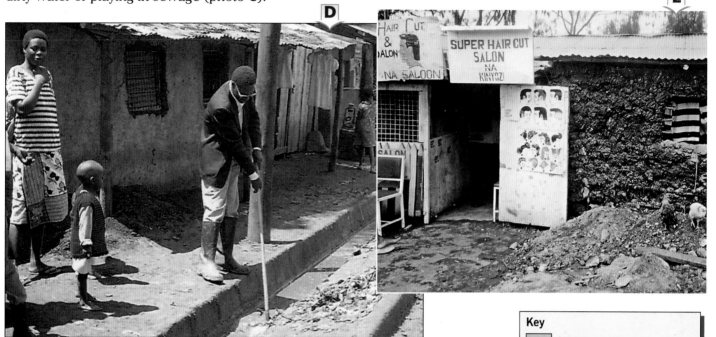

D

E

Activities

1 Give three differences between photos **A** and **B**.

2 Map **F** shows the location of shanty settlements in Nairobi.

   a How far is Kibera from the city centre?

   b Which direction is Kibera from the city centre?

   c Write out the following sentence using the correct word from the pair in brackets.

   > Shanty settlements are areas of (good/poor) quality housing found at the (edge/middle) of the city on (good/poor) quality building land.

3 Using photo **C**, list four problems of life in a shanty settlement.

4 List four ways that people living in a shanty settlement can earn money.

F

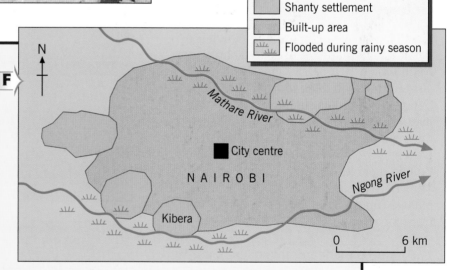

Key
- Shanty settlement
- Built-up area
- Flooded during rainy season

N

Mathare River

City centre

NAIROBI

Ngong River

Kibera

0   6 km

5 **a** List four good points about living in Nairobi.

   b Why is Nairobi described as having 'two sides'?

Summary

Cities in developing countries have two sides. Well-off people live and work in good conditions near to the city centre. Poor people often live and work in less pleasant shanty settlements a long way from the city centre.

What is the Maasai way of life?

The Maasai in Kenya

One ethnic group living in Kenya are the Maasai (see map **A** on page 112). They are **pastoralists** with herds of cattle and goats. Some are **nomadic** and have to move about to find water and grass for their animals. The Maasai depend on these animals for their daily food. It is cattle, not money, which means wealth to the Maasai. Indeed, the usual Maasai greeting is 'I hope your cows are well'.

The land where the Maasai live is fairly flat and covered with grass which depends upon the rain. The 'long rains' come between April and June, the 'short rains' in October and November. In the dry months the grass withers under the hot sun. If the rains do not come then the Maasai have to move to look for grass for their animals.

Houses

Most Maasai live in an **enkang** which is a small village made up of 20 to 50 huts (photo **A**), in which 10 to 20 families live. It is surrounded by a thick thorn hedge to keep out dangerous animals such as lions, leopards and hyenas. Tiny passages allow people and, in the evening, cattle to pass through. These passages are blocked up at night. The huts are built in a circle around an open central area. They barely reach the height of an adult Maasai (photo **B** and plan **D**) and are built from local materials.

The frame is made from wooden poles. Mud, from nearby rivers, and cow dung are used for the walls. Grass from the surrounding area is used for the roof. The hut is entered by a narrow tunnel. Apart from an opening the size of a brick, there are no windows or chimneys. The inside is dark and full of smoke from the fire. It is cool during the day, warm at night and free from flies and mosquitoes. Cowskins are laid on the floor for beds. Water and honey are stored in gourds – a ball-shaped plant with a thick skin.

A

B

C

Plan of the inside of a Maasai hut

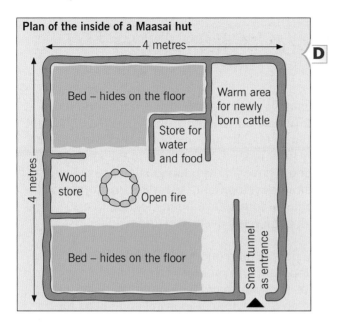

- 4 metres
- 4 metres
- Bed – hides on the floor
- Warm area for newly born cattle
- Store for water and food
- Wood store
- Open fire
- Bed – hides on the floor
- Small tunnel as entrance

D

E

Dress

Men wear brightly coloured 'blankets' and women wear lengths of cloth (photo **A**). The women shave their heads and wear beads around their neck, wrists and ankles. Teeth are cleaned with sticks and the men also use sticks to comb their hair. As water is often scarce, the Maasai use animal and vegetable fat to clean themselves and sweet smelling grasses for perfume.

Daily jobs

The women have to collect sticks for the fire and water for cooking. They also make baskets and jewellery. The men spend the whole day guarding their animals. The Maasai regard the ground as sacred and believe it should not be broken. This means no crops can be grown, wells cannot be dug and, often, the dead are not buried but are left to wild animals. As crops are not grown sometimes animals are exchanged for grain.

Diet

A major part of the Maasai diet is milk mixed with blood from their cows. In times of drought only blood is used.

Activities

1. **a** Give two facts about a developing country.
   b What is meant by an 'ethnic group'?
   c In which continent is Kenya?

2. **a** Give two reasons why cattle are important to the Maasai.
   b Write a paragraph to describe Maasai farming. Include these words:
   - cattle and goats
   - flat land
   - grass
   - rain
   - nomadic

3. Sketch **F** shows a Maasai hut and several questions. Draw the hut and add labels by answering the questions.

4. How do photo **A** and your drawing of a Maasai hut suggest that the weather is:
   a usually warm   **b** not very wet?

F

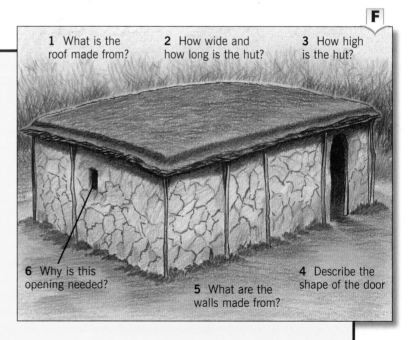

1 What is the roof made from?

2 How wide and how long is the hut?

3 How high is the hut?

6 Why is this opening needed?

5 What are the walls made from?

4 Describe the shape of the door

Summary

Landscape, weather and wealth all affect the family life, housing, clothing and diet of the Maasai.

Can development in Kenya be sustainable?

Kenya became an independent country in 1963. Since then it has become one of the more successful **developing countries** in East Africa. Economic growth has brought many benefits to Kenya. Some parts of the country are now developed and many Kenyans are wealthy and enjoy a high standard of living.

A

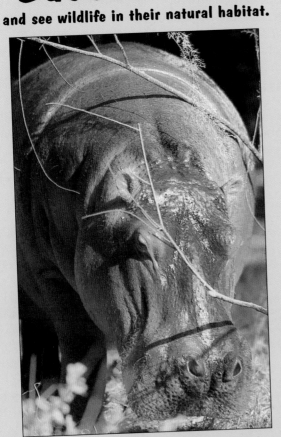

Visit

Baobab Farm
and see wildlife in their natural habitat.

* Wild animals including hippopotamus, elephant, giraffe, giant tortoise, wildebeest and zebra.
* Fish farm and crocodile hatchery.
* Reptile house with snakes and insect displays.
* Aviary with spectacular display of tropical birds.
* Guided nature trails and educational facilities.
* Information centre, shops and restaurant.

Unfortunately, progress has also brought problems both for people and the environment. Wealth has not been shared equally and the majority of Kenyans are still very poor indeed. Many living in the **shanty settlements** have an even lower standard of living than before development. As for the environment, large areas of forest have been cleared, rivers polluted, coral reefs destroyed and wildlife lost.

Some people are concerned that so-called development and progress is actually damaging Kenya. They say that development should bring about an improvement in living conditions for people. At the same time, however, it should not harm or destroy the environment nor prevent people in the future from achieving similar standards of living. Development like this is called **sustainable development**.

There are many examples in Kenya of sustainable development projects. One of these is at Baobab Farm just 10 km north of Mombasa on the coast of Kenya. The area was once a huge quarry where 25 million tonnes of coral limestone had been removed, leaving a huge, ugly scar. As photo **B** shows, the environment had been totally destroyed – there was no soil, no vegetation and no wildlife.

In 1971, the cement company that owned the quarry decided to try and restore the area to its original state. They planted quick-growing casuarina trees (photo **C**) and introduced thousands of red millipedes to break down the fallen leaves into **humus**. Earthworms, grasses and more trees were added but no chemicals were used. Within 20 years sufficient soil had formed to allow a tropical rainforest environment to develop (photo **D**).

The project is not just an environmental success. It has also become a sustainable commercial venture. Baobab Farm, as it is called, is open to school parties every morning, while in the afternoon it attracts other visitors. It is now one of Kenya's most visited tourist attractions.

B Bamburi limestone quarry

Before

C Casuarina trees planted

During

D Baobab Farm restored environment

After

Activities

1 Complete these three sentences:
   a Sustainable development is ...
   b Sustainable development should ...
   c Sustainable development should not ...

2 Look at the following list of statements about the restoration scheme at Baobab Farm. Copy and complete diagram **E** by putting each statement into the correct box.
   - Casuarina trees planted
   - Tropical rainforest returns
   - Tourist facility opened
   - Limestone quarried
   - Red millipedes introduced
   - Bare, ugly landscape
   - New soil develops
   - No soil, vegetation or wildlife
   - Wildlife introduced

E

Before
* ....
* ....
* ....

⬇

During
* ....
* ....
* ....

⬇

After
* ....
* ....
* ....

3 Draw a star diagram to show the main attractions of Baobab Farm.

4 What features of Baobab Farm make it a good example of sustainable development?

Summary

Sustainable development should improve people's quality of life and standard of living without wasting resources or destroying the environment. Kenya's Baobab Farm is an example of successful sustainable development.

What is a developing country?

By now you should be aware of many differences between living in Kenya and in your home region in the UK. These differences include ethnic groups, dress, housing, jobs, wealth and the quality of life. Kenya is an example of a **developing country**. What is a developing country? How is life in a developing country different from life in a developed country like the UK?

In the UK most, but not all, people earn a lot of money compared with those in a developing country. They live in good houses, have their own cars and videos and can afford good food and holidays. Compared with Kenya most people in Britain have a high **standard of living**. Kenya is considered to be 'poor' and the UK to be 'rich'. Most people see the difference in wealth as the main difference between a developing country and a developed country.

The wealth of a country is given by its **gross national product (GNP)**. This is the total amount of money made by a country from its raw materials, its manufactured goods and its services.

Notice that GNP is always given in American dollars (US$). The total amount can then be divided by the total number of people living in that country. This gives the average amount of money available for every person living in the country.

By giving the GNP in US$ it is easy to compare different countries. Table **A** gives the average income (GNP) per person for five developing countries and three developed countries.

Apart from wealth there are many other ways of trying to measure the level of a country's development (table **B**).

A

	Country	GNP (US$ per person)
Developing countries	Bangladesh	2,000
	Brazil	8,100
	Egypt	4,200
	Kenya	1,100
	Peru	5,600
Developed countries	Japan	29,400
	UK	29,600
	USA	40,100

B

Jobs		Primary activities give most jobs in a developing country. A developed country has fewer primary activities and more secondary and service jobs.
Trade		A developing country usually has to sell raw materials at a low price and has to buy manufactured goods from developed countries at a high price.
Population		A developing country has a higher **birth** and **death rate** (page 94), more young children dying (high **infant mortality**), adults dying at a younger age (short **life expectancy**) and a faster population increase than a developed country.
Health		A developing country has less money to spend on training doctors and nurses and in providing hospitals and medicines.
Education		The large number of children and lack of money for schools in a developing country mean fewer people can read and write than in a developed country (low **literacy rate**).

The 'rich' North and the 'poor' South

What happens when the different methods used in table **B** to measure the level of development of a country are put together? The result is a group of mainly 'rich' countries which are found in the North and a group of mainly 'poor' countries which lie to the South. Map **C** shows this division.

C

North. Most countries north of the line are said to be **developed**.

The 'rich' North
- Large GNP
- Most jobs in industry and services
- Large amounts of trade
- Many doctors and hospitals
- Low birth and death rates
- Few children die young
- Slow population growth
- Long life expectancy
- Many schools

The 'poor' South
- Small GNP
- Most jobs in farming
- Small amounts of trade
- Few doctors and hospitals
- High birth and death rates
- Many children die young
- Rapid population growth
- Short life expectancy
- Few schools

South. Most countries south of the line are said to be **developing**.

Activities

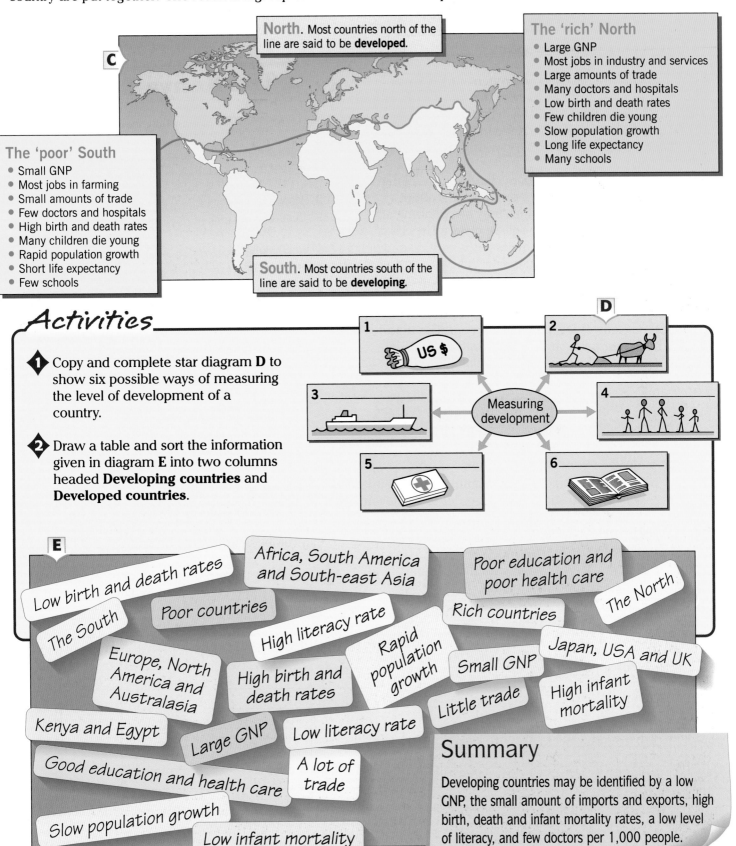

1 Copy and complete star diagram **D** to show six possible ways of measuring the level of development of a country.

2 Draw a table and sort the information given in diagram **E** into two columns headed **Developing countries** and **Developed countries**.

D

1 ____ US $

2 ____

3 ____

Measuring development

4 ____

5 ____

6 ____

E

Low birth and death rates

Africa, South America and South-east Asia

Poor education and poor health care

The South

Poor countries

High literacy rate

Rich countries

The North

Europe, North America and Australasia

High birth and death rates

Rapid population growth

Small GNP

Japan, USA and UK

Kenya and Egypt

Large GNP

Low literacy rate

Little trade

High infant mortality

Good education and health care

A lot of trade

Slow population growth

Low infant mortality

Summary

Developing countries may be identified by a low GNP, the small amount of imports and exports, high birth, death and infant mortality rates, a low level of literacy, and few doctors per 1,000 people.

121

We have already seen that all countries are different. Some are rich and have high standards of living whilst others are poor and have low standards of living. Countries that differ in this way are said to be at different stages of development.

In this unit you have learned much about Kenya. You will certainly know by now that Kenya is very different from the UK and also, certainly in terms of wealth, that it is very poor. Indeed Kenya is an almost perfect example of a **developing country**.

As we have seen, however, measuring development can be difficult. Development, after all, is about quality of life. Some countries may be economically poor, but their people can still be cheerful, relaxed and generally happy with their lives.

In this enquiry you work for a department of the British Government responsible for overseas development and have been asked to make a report on Kenya's level of development. The report will be in four parts, as shown in drawing **A**. Pages 120 and 121 of this book will be helpful to you.

How developed is Kenya?

 a Make a copy of table **B** below.

b Using information from diagram **C**, complete your table to show the rank order of the seven countries for each measure of development. The most developed will score 1 and the least developed will score 7.

c Add the scores together for each country, and complete the Total column.

 A

 **British Overseas Aid**

The Kenya Report

1 How developed is Kenya compared with the UK?

2 How developed is Kenya compared with neighbouring countries in Africa?

3 How developed is Kenya in terms of social and cultural measures of development?

4 In which areas of development (as shown in table **B**) is Kenya in greatest need of improvement? What might be done to help the country make progress?

In each case suggest reasons for your answer. You should consider economic, social and cultural factors.

- **Economic factors** are about the wealth of a country.
- **Social factors** are concerned with standards of living and quality of life.
- **Cultural factors** are about traditions and the way of life.

2 Write out the countries from your completed table as a 'League table of development'. The country with the lowest score will be most developed and should be at the top of the league.

3 Complete the report using your answer to activity 1 and information in diagrams **C** and **D**.

	Country	Wealth (highest first)	Trade (highest first)	Life expectancy (highest first)	People/doctor (lowest first)	Literacy rate (highest first)	TOTAL
B	UK						
	Ethiopia						
	Kenya						
	Somali Rep.						
	Sudan						
	Tanzania						
	Uganda						

C

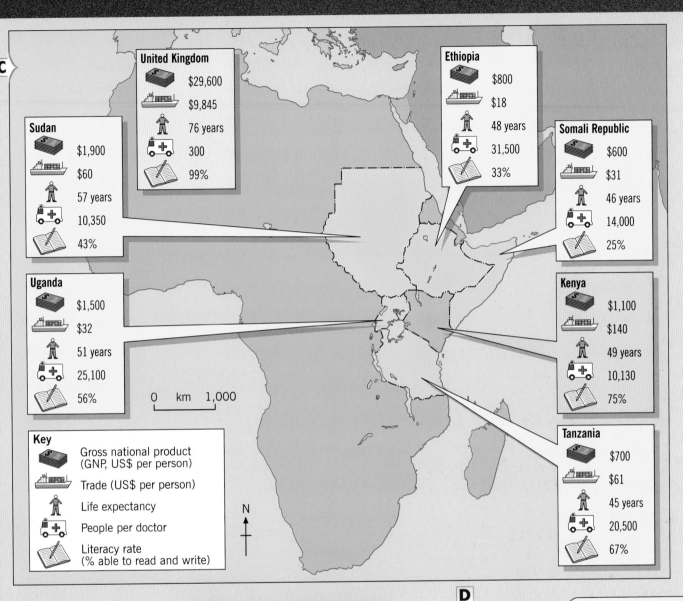

United Kingdom
- $29,600
- $9,845
- 76 years
- 300
- 99%

Ethiopia
- $800
- $18
- 48 years
- 31,500
- 33%

Sudan
- $1,900
- $60
- 57 years
- 10,350
- 43%

Somali Republic
- $600
- $31
- 46 years
- 14,000
- 25%

Uganda
- $1,500
- $32
- 51 years
- 25,100
- 56%

Kenya
- $1,100
- $140
- 49 years
- 10,130
- 75%

Tanzania
- $700
- $61
- 45 years
- 20,500
- 67%

Key
- Gross national product (GNP, US$ per person)
- Trade (US$ per person)
- Life expectancy
- People per doctor
- Literacy rate (% able to read and write)

0   km   1,000

N

D

Our country is making progress but most of us are still very poor.

We have a wealth of beautiful scenery and exciting wildlife.

Family values are important to us and we are always willing to help others.

We still need a clean, reliable water supply and more food.

Kenyan children always seem to have a smile on their faces and be full of fun.

Our traditional way of life is interesting and colourful.

We have some of the best athletes in the world.

People in Kenya are well known for being cheerful and friendly.

What are world issues?

What is this unit about?

This unit looks at some of the major problems facing our world. These problems are called world issues because they affect the lives of people across the world. Such issues may only be solved by international co-operation.

In this unit you will learn about:

◆ the causes and effects of global warming

◆ how our use of energy may change

◆ the problems of food supply and water shortages

◆ the poverty problem.

A London

Why is it important to know about world issues?

One very good reason is that some issues are likely to affect you personally. For example, global warming will affect the way you live, having too much food will affect your health, and requests for aid will give you the choice of helping others.

But there is much more to it than that. Learning about the problems facing our world can help you appreciate the need to look after our planet. It can help you become a global citizen, interested in the state of the world, aware of the problems and willing to do your bit to solve them.

◆ Look at photos **A** and **B**. For each one:
  – describe what has happened
  – describe the problems
  – suggest what caused the problems
  – suggest what might be done to reduce the problems.

◆ How would you feel about living in place **B**?

◆ List the problems that the family in photo **C** may have.

B India

What is global warming?

The world is now warmer than it has been for many thousands of years. In the last century, average temperatures rose by almost 1°C, with the greatest increases in the last 40 years. This century has already seen the warmest years ever recorded. This heating up of our planet is called **global warming** and will cause serious problems in the future.

Global warming is thought to be due to the **greenhouse effect**. As diagram **B** shows, the earth is surrounded by a layer of gases including carbon dioxide. These keep the earth warm by preventing the escape of heat that would normally be lost to the atmosphere. The gases act rather like the glass in a greenhouse. They let heat in but prevent most of it from getting out (diagram **A**).

The burning of fossil fuels such as coal, oil and natural gas produces large amounts of carbon dioxide, which, as diagram **C** shows, is the main greenhouse gas. As the amount of this gas increases, the earth becomes warmer.

Most scientists believe that over the next hundred years temperatures will rise by 2 to 4°C. Some think it may be even more than that but a few believe that other effects might cancel out this warming. Whoever is right, it is fairly certain that in the years ahead it is unlikely that conditions on earth will be the same as they are now.

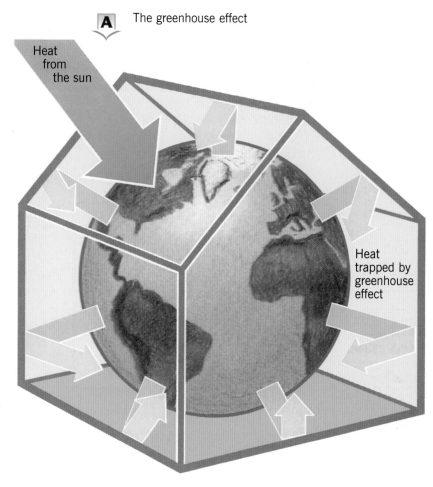

A The greenhouse effect

Heat from the sun

Heat trapped by greenhouse effect

B Causes of the greenhouse effect

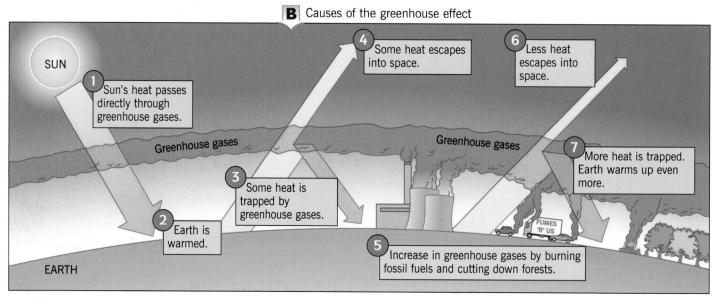

SUN

1 Sun's heat passes directly through greenhouse gases.

Greenhouse gases

2 Earth is warmed.

3 Some heat is trapped by greenhouse gases.

4 Some heat escapes into space.

5 Increase in greenhouse gases by burning fossil fuels and cutting down forests.

Greenhouse gases

6 Less heat escapes into space.

7 More heat is trapped. Earth warms up even more.

FUMES 'R' US

EARTH

Global warming is a world problem. Almost every country contributes in some way to producing greenhouse gases. Global warming affects every corner of the planet.

International agreement is needed if we are to reduce greenhouse gases and slow down global warming. So far, this agreement has been difficult to achieve and global warming continues to be a real problem for our world.

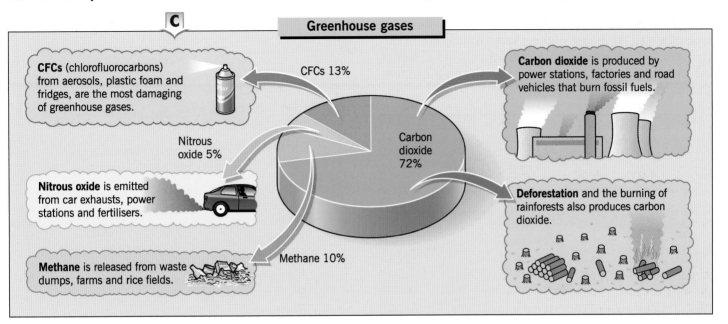

C

Greenhouse gases

CFCs (chlorofluorocarbons) from aerosols, plastic foam and fridges, are the most damaging of greenhouse gases.

CFCs 13%

Carbon dioxide is produced by power stations, factories and road vehicles that burn fossil fuels.

Nitrous oxide 5%

Carbon dioxide 72%

Nitrous oxide is emitted from car exhausts, power stations and fertilisers.

Deforestation and the burning of rainforests also produces carbon dioxide.

Methane is released from waste dumps, farms and rice fields.

Methane 10%

Activities

1 Look at graph **D**.

a Describe the change in global temperatures.

b Describe the change in carbon dioxide levels.

c Explain the link between the two.

2 a What is global warming?

b What is the greenhouse effect?

c Give two reasons why global warming can be described as a world problem.

3 a Make a larger copy of diagram **E**.

b Add labels to your diagram to explain how the burning of fossil fuels may cause global warming.

4 a Which greenhouse gases result from burning coal, oil or natural gas?

b Why is international agreement needed to reduce greenhouse gases?

c Make a list of ways in which you may have contributed to global warming in the last week.

Global temperatures and carbon emissions

D

Temperature (°C)

Carbon dioxide (billion tonnes)

Temperature

Carbon dioxide

14

13

400

300

200

100

0

1860 1880 1900 1920 1940 1960 1980 2000

Year

E

Summary

The burning of fossil fuels has caused an increase in temperatures around the world. This is called global warming.

What are the effects of global warming?

Nobody knows exactly what the effects of global warming will be. Some of the effects will no doubt be harmful, but others may bring benefits. Some of the effects predicted by scientists are shown on these two pages.

A

B How is global warming changing our world?

- Sea temperatures are rising, causing the water to expand and the sea level to rise.
- Ice caps and glaciers are melting, causing sea levels to rise even further.
- Climates around the world are changing. Some are getting more rain, some less.

How are these changes likely to affect us?

- Low-lying coastal areas will be flooded. Some islands will disappear altogether.
- Places with less rain may experience food shortages as their crops fail to grow.
- Plants and animals that cannot adapt to climate change will become extinct.
- There might be an increase in insect pests.
- Tropical diseases may spread to temperate regions like the UK.

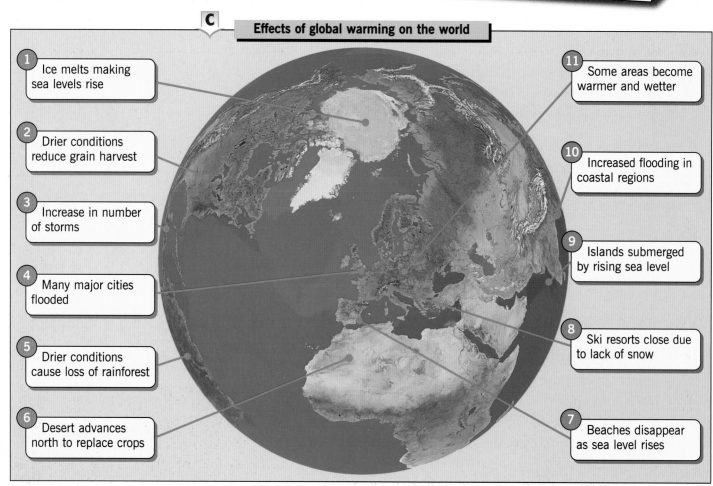

C Effects of global warming on the world

1 Ice melts making sea levels rise

2 Drier conditions reduce grain harvest

3 Increase in number of storms

4 Many major cities flooded

5 Drier conditions cause loss of rainforest

6 Desert advances north to replace crops

7 Beaches disappear as sea level rises

8 Ski resorts close due to lack of snow

9 Islands submerged by rising sea level

10 Increased flooding in coastal regions

11 Some areas become warmer and wetter

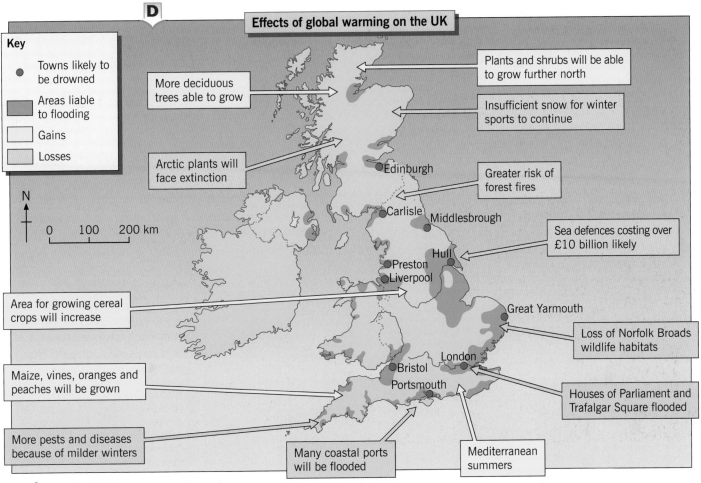

D

Effects of global warming on the UK

Key

- Towns likely to be drowned
- Areas liable to flooding
- Gains
- Losses

N
0   100   200 km

Plants and shrubs will be able to grow further north

More deciduous trees able to grow

Insufficient snow for winter sports to continue

Arctic plants will face extinction

Greater risk of forest fires

Edinburgh

Carlisle

Middlesbrough

Sea defences costing over £10 billion likely

Hull

Preston
Liverpool

Area for growing cereal crops will increase

Great Yarmouth

Loss of Norfolk Broads wildlife habitats

London

Maize, vines, oranges and peaches will be grown

Bristol
Portsmouth

Houses of Parliament and Trafalgar Square flooded

More pests and diseases because of milder winters

Many coastal ports will be flooded

Mediterranean summers

Activities

1 Look at satellite photo **C**. Match the locations on the photo with the places below. The world map on the back cover will help you.

- Caribbean Sea • Sahara Desert • Arctic Ocean
- Amazon • Mediterranean Sea • Bangladesh
- Maldives • Europe • Great Plains • Alps • London

2 a Make a larger copy of diagram **E**.

b Put the statements in the panel below into the correct boxes coloured yellow.

c Complete the 'Effects' boxes (coloured green) with six examples from table **B** or photo **C**.

- Water expands • Sea levels rise • Ice cap melts
- Plants and wildlife affected • Climate changes

3 Describe how global warming in the UK:

a will bring benefits to farmers

b will cause most problems for people

c will affect the area where you live.

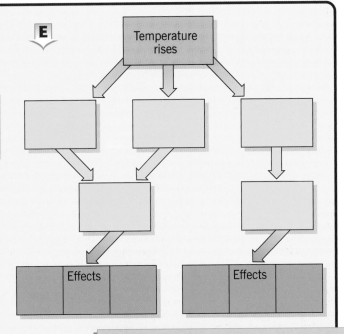

E

Temperature rises

Effects   Effects

Summary

Global warming will cause serious problems throughout the world due mainly to rising sea levels and climate change. Some of the effects may bring a number of benefits.

How can our energy use change?

Fossil fuels are important to us. Coal, oil and natural gas provide almost all of our energy needs and are essential to modern-day living. They are used to generate electricity, heat our homes and power planes, cars and other vehicles.

But fossil fuels are running out and as they are non-renewable, cannot be replaced. We need to find alternative forms of energy and in the meantime, carefully conserve those resources that we have.

When will we run out?
At our present rate of use ...
◆ Oil ... in about 50 years
◆ Gas ... in about 60 years
◆ Coal ... in about 250 years

Many experts believe that hydrogen, wind and solar power will provide most of the world's energy in the future. The drawing below shows how this may happen.

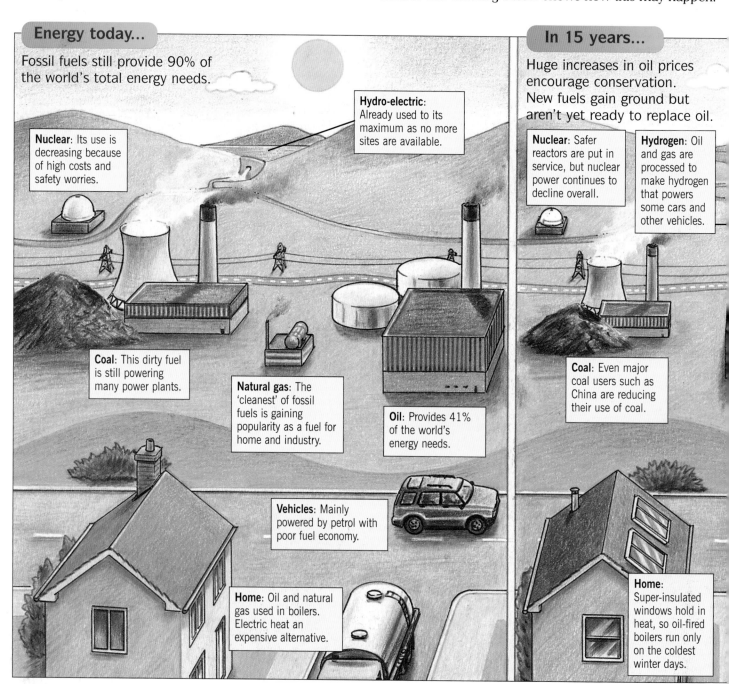

Energy today...

Fossil fuels still provide 90% of the world's total energy needs.

Hydro-electric: Already used to its maximum as no more sites are available.

Nuclear: Its use is decreasing because of high costs and safety worries.

Coal: This dirty fuel is still powering many power plants.

Natural gas: The 'cleanest' of fossil fuels is gaining popularity as a fuel for home and industry.

Oil: Provides 41% of the world's energy needs.

Vehicles: Mainly powered by petrol with poor fuel economy.

Home: Oil and natural gas used in boilers. Electric heat an expensive alternative.

In 15 years...

Huge increases in oil prices encourage conservation. New fuels gain ground but aren't yet ready to replace oil.

Nuclear: Safer reactors are put in service, but nuclear power continues to decline overall.

Hydrogen: Oil and gas are processed to make hydrogen that powers some cars and other vehicles.

Coal: Even major coal users such as China are reducing their use of coal.

Home: Super-insulated windows hold in heat, so oil-fired boilers run only on the coldest winter days.

Activities

1 a Which energy resource is most used today? Give a list of uses.

b Which energy resource is likely to be most used in 50 years' time? Give a list of uses.

2 During the next 50 years, which energy sources are expected to:

a be used less **b** be used more?

3 How will changes in energy use in the next 50 years affect global warming? Give reasons for your answers.

4 How would it affect you if oil ran out? Think about your day-to-day life.

5 In what ways do you think the energy sources that may be used in 50 years' time are better than the ones we use today?

Summary

There are limited supplies of fossil fuels and once they are gone, they cannot be replaced. Other forms of energy need to be developed in their place.

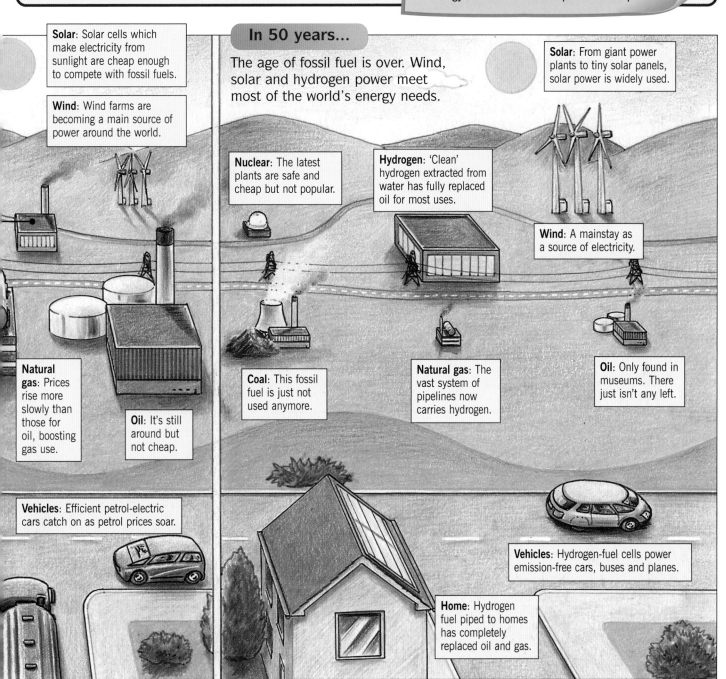

Solar: Solar cells which make electricity from sunlight are cheap enough to compete with fossil fuels.

Wind: Wind farms are becoming a main source of power around the world.

Natural gas: Prices rise more slowly than those for oil, boosting gas use.

Oil: It's still around but not cheap.

Vehicles: Efficient petrol-electric cars catch on as petrol prices soar.

In 50 years...

The age of fossil fuel is over. Wind, solar and hydrogen power meet most of the world's energy needs.

Nuclear: The latest plants are safe and cheap but not popular.

Hydrogen: 'Clean' hydrogen extracted from water has fully replaced oil for most uses.

Solar: From giant power plants to tiny solar panels, solar power is widely used.

Wind: A mainstay as a source of electricity.

Coal: This fossil fuel is just not used anymore.

Natural gas: The vast system of pipelines now carries hydrogen.

Oil: Only found in museums. There just isn't any left.

Home: Hydrogen fuel piped to homes has completely replaced oil and gas.

Vehicles: Hydrogen-fuel cells power emission-free cars, buses and planes.

131

What is the water problem?

In Britain we take it for granted that, apart from the occasional drought, a reliable supply of clean water is always there when we need it. This is because:

◆ we get rain spread fairly evenly throughout the year

◆ we have the money and technology to create reservoirs from which water can be piped to our homes and places of work and leisure.

This is not the case for many people in other parts of the world, especially those living in developing countries. The United Nations claim that two out of every five of the earth's population lack a safe and reliable supply. They also suggest that there is enough fresh water to support five times the present world's population. However, this water is either unevenly spread or is hard to reach.

As the demand for water grows, this normally renewable resource is increasingly under pressure. Diagram **A** gives some reasons why developing countries may be short of water. Diagram **D** shows some of the effects of water shortages.

A

Causes

Rural–urban migration to large towns and cities. As the population grows then more water is needed.

Global warming means that many parts of the world will get less rainfall.

Water shortages occur in poorer countries as they have neither the money nor the technology to create reservoirs or to lay water pipes.

Wells can dry up as more water is used.

Not enough rain can cause a drought.

Rivers may be used for drinking water, washing in and for getting rid of sewage.

Many towns and cities in less developed countries have their limited water supply polluted due to poor sewerage and hygiene. Up to 40 families may have to share one tap.

Activities

1 Look at diagrams **A** and **B**.

   a Give reasons why so many people are short of water.

   b Why are water shortages greatest in developing countries?

2 Look at diagram **D**.

   a How many people are likely to be short of water in 2050?

   b Why will it be difficult to provide water for so many people?

B

UNITED NATIONS

UN Report 2005

● Water-related diseases claim over 2.2 million lives a year – that is one death every 15 seconds.

The United Nations predict that by 2050 the number of people who will be short of fresh water will rise to over three in every five. They also predict that water shortages may cause problems between countries that, at present, rely upon the same water supply.

There are no easy solutions to the world's water shortage. Building dams to create reservoirs is out of favour. Many people in the poorest of developing countries have to rely upon help from voluntary organisations like WaterAid (diagram **C**). WaterAid collects money through donations in the UK. It then uses this money either to provide clean water or to improve sanitation in poor countries.

C

25 YEARS 1981-2006
WaterAid

◆ The UK's only charity dedicated solely to providing safe, clean water, sanitation and hygiene education to the world's poorest people. It believes these basic services to be essential to life. Without them vulnerable communities have little chance of escaping the stranglehold of disease and poverty.

◆ So far WaterAid has helped 8.5 million people. Its aim is to help beat the threat of death and misery caused by a lack of clean water and give people hope for a better future.

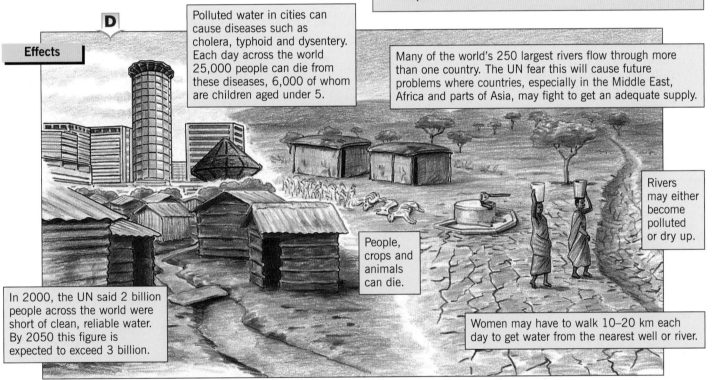

D

Effects

Polluted water in cities can cause diseases such as cholera, typhoid and dysentery. Each day across the world 25,000 people can die from these diseases, 6,000 of whom are children aged under 5.

Many of the world's 250 largest rivers flow through more than one country. The UN fear this will cause future problems where countries, especially in the Middle East, Africa and parts of Asia, may fight to get an adequate supply.

Rivers may either become polluted or dry up.

People, crops and animals can die.

In 2000, the UN said 2 billion people across the world were short of clean, reliable water. By 2050 this figure is expected to exceed 3 billion.

Women may have to walk 10–20 km each day to get water from the nearest well or river.

3 Look at the statistics on the back cover.

  a List the countries in order of:
- wealth
- access to clean water.

  b Write a sentence to describe the link between the wealth of a country and its access to clean water. Suggest reasons for this.

4 **a** Photo **E** shows clean water provided by WaterAid in Nepal. How can organisations like WaterAid help in providing clean water to people living in developing countries?

  b How might **you** be able to help?

E

Summary

Nearly half of the world's people have no reliable supply of clean water. The situation is expected to get worse.

Food – too little or too much?

Many of us living in a developed country such as the UK are used to having at least three good meals a day. Added to that are various snacks that we take any time we are hungry, thirsty or, in some cases, just bored. In contrast, many people who live in poorer, developing countries consider themselves lucky if they get one good meal a day.

For people to have a satisfactory diet they need:

◆ the correct **quantity** of food – the amount of food a person eats is measured in calories

◆ the correct type or **quality** of food – a healthy diet consists of proteins (meat, eggs and milk), carbohydrates (cereals and potatoes) and vitamins (fruit and vegetables, meat and fish).

Diet can be a problem in both poor and rich countries.

The United Nations say that there is enough food produced each year to feed everybody in the world. Unfortunately while rich countries like the USA and those in Western Europe produce more than they need, many poorer countries, especially in Africa, do not produce enough.

A

The United Nations

◆ The UN claims that 600 million people were short of food in the year 2000.

◆ It predicts that this number will have risen to 3,300 million by 2050.

Some causes of food shortages

B

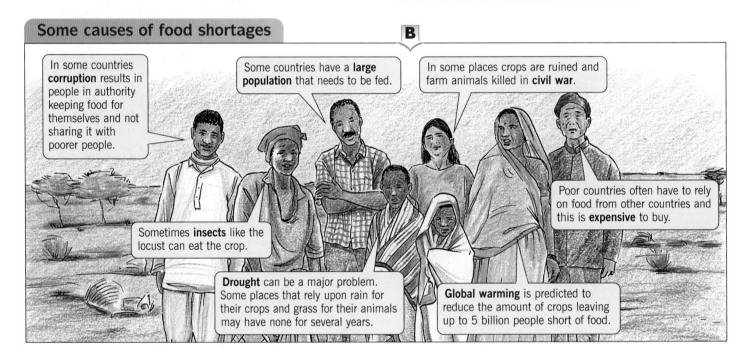

In some countries **corruption** results in people in authority keeping food for themselves and not sharing it with poorer people.

Some countries have a **large population** that needs to be fed.

In some places crops are ruined and farm animals killed in **civil war**.

Poor countries often have to rely on food from other countries and this is **expensive** to buy.

Sometimes **insects** like the locust can eat the crop.

Drought can be a major problem. Some places that rely upon rain for their crops and grass for their animals may have none for several years.

Global warming is predicted to reduce the amount of crops leaving up to 5 billion people short of food.

Poor countries

By 2050 half of the world's population is likely to be underfed (diagram **A**). Diagram **B** gives several reasons why certain places are short of food. A person who does not get enough to eat or the right type of food is likely to suffer from **malnutrition**. Malnutrition usually results from poverty when people have not got enough money to buy food, rather than because there is too little food for the number of people. Diagram **C** shows some of the effects of malnutrition.

Rich countries

A person who either eats too much or who eats the wrong type of food is also unhealthy. An increasing number of people in developed countries like the USA and the UK are overweight. They are said to be **obese**. Diagram **C** also shows some of the effects of being overweight. In the UK:

◆ what we like to eat is not always good for us

◆ what we should eat we often do not like.

Your school may be one of those that has been encouraged by the television and the government to change its dinner menus to provide a healthier diet.

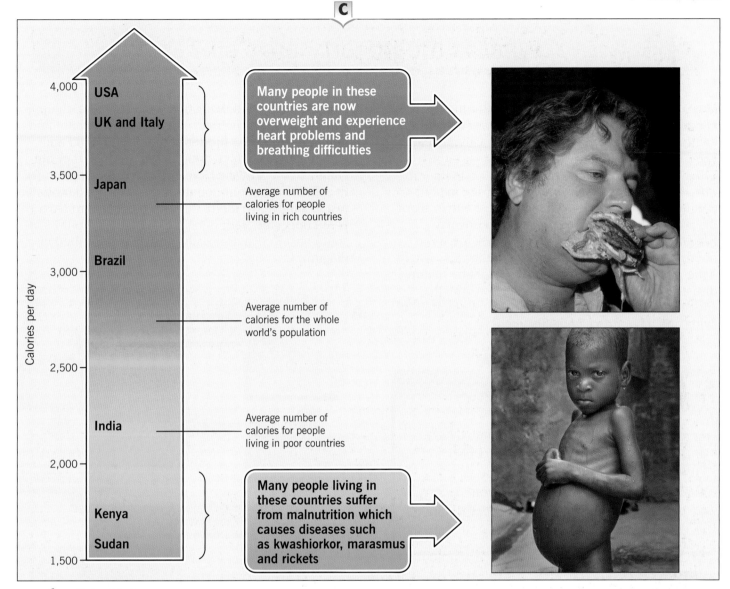

Calories per day

4,000 — USA

UK and Italy

Many people in these countries are now overweight and experience heart problems and breathing difficulties

3,500 — Japan

Average number of calories for people living in rich countries

Brazil

3,000 —

Average number of calories for the whole world's population

2,500 —

India

Average number of calories for people living in poor countries

2,000 —

Kenya

Many people living in these countries suffer from malnutrition which causes diseases such as kwashiorkor, marasmus and rickets

Sudan

1,500 —

Activities

1 Using diagram **B**, list the causes of food shortages under these headings:

Natural causes | Caused by rich countries

Caused by poor countries

2 a Describe what would be a good diet for yourself. Include the number of calories and the type of food that would give you a healthy diet.

b What changes to your present diet would help you and your family to live a healthier life?

3 a Look at the statistics on the back cover. On a copy of table **D**, list in rank order, with the highest first, the percentage of people in the six countries suffering from
  • obesity • malnutrition.

b What is obesity?

c What are the causes and effects of obesity?

d What is malnutrition?

e What are the causes and effects of malnutrition?

	% obese	% malnourished
1		
2		
3		

D

Summary

There are considerable differences in the quantity and quality of food supplies between rich and poor countries. A poor diet can cause many illnesses.

What is the poverty problem?

Most of us in the UK live in houses with running water, sewerage and electricity. We also tend to take for granted other basic needs such as education, health care, jobs and access to plenty of food.

This is not the case for many people living in poor countries. It is now believed that almost one in six of the world's population live in **extreme poverty** (diagram **A**). These people not only lack the basic needs available in rich countries, but they also live in difficult environments where there might be natural disasters, civil war and rapid population growth. Together this means that they have a low standard of living and a poor quality of life.

Poor countries find that it is impossible to catch up with the richer countries. Indeed, many of them find that they are falling further and further behind and getting increasingly poorer. This is because they are caught in the so-called **cycle of poverty** (diagram **B**). People living in poor countries often face a daily battle just to survive. To them, the future appears to have little hope. One such person is Marietta, who lives in south-east Kenya. Her daily lifestyle is described in diagram **C**.

A

GLOBAL NEWS
World Poverty in 2005

- 800 million people live in extreme poverty.
- They earn less than US$1 a day, which is about £200 a year.
- 30,000 children die each year from poverty.
- 23 of the world's 25 poorest countries are in Africa.

B

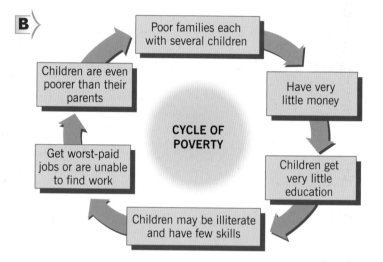

CYCLE OF POVERTY

Poor families each with several children → Have very little money → Children get very little education → Children may be illiterate and have few skills → Get worst-paid jobs or are unable to find work → Children are even poorer than their parents →

C

Marietta lives with her seven children, all aged under 12 years. Her husband works 250 km away in Mombasa but he never earns enough either to travel home or to send any money. With her nearest neighbour living 3 km away, Marietta is left alone to look after her small shamba (farm).

Her day begins as soon as it is light. She collects wood as this is the only source of energy for both cooking and warmth. The eldest girls have to go 2 km to the nearest river to collect the day's supply of water before school. They then walk another 5 km to school (there is no transport). Marietta spends most of her day collecting firewood and looking after her crops of maize, beans and sorghum. She also keeps 12 chickens, 20 goats and two cows. The cows are her main source of wealth. They provide milk and are used to plough the hard, dry ground.

It is essential that they remain healthy as the nearest vet is over 50 km away and his bill would be too much for Marietta to pay.

Some causes and effects of poverty

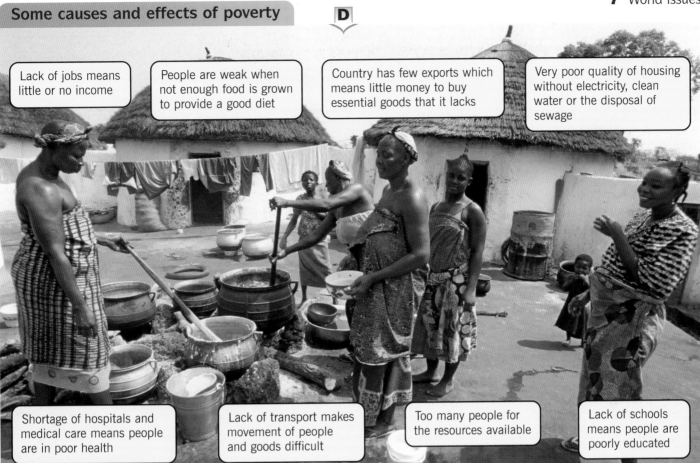

Lack of jobs means little or no income

People are weak when not enough food is grown to provide a good diet

Country has few exports which means little money to buy essential goods that it lacks

Very poor quality of housing without electricity, clean water or the disposal of sewage

Shortage of hospitals and medical care means people are in poor health

Lack of transport makes movement of people and goods difficult

Too many people for the resources available

Lack of schools means people are poorly educated

Activities

1 a What is meant by 'The cycle of poverty'?

b Make a larger copy of diagram **E**. Put the following into the correct boxes to show the effects of the cycle of poverty.
- Family becomes even poorer
- Family cannot afford to visit the very few doctors in the country
- Family become weaker and are not well enough to work
- Family likely to fall ill

c Describe how Marietta is caught in the cycle of poverty.

d How is daily life for Marietta and her daughters different from that of your family?

2 Look at the statistics on the back cover.

a Name the two poorest countries.

b With the help of diagrams **D** and **F**, list seven problems caused by poverty in these countries.

c Describe any five of these problems.

E

Family has little money for food

F

Jobs   Housing   Trade

Resources → **Causes of poverty** ← Transport

Education and health   Food supply

Summary

Poverty is a problem that affects large numbers of people around the world. It is difficult for people living in these countries to break out of the cycle of poverty.

How might poverty be reduced?

A major problem facing the world today is how to reduce poverty in poor countries and how to improve the standard of living and quality of life of people living there. This means trying to find ways to help people living there to break free from the cycle of poverty described on page 136. This can only be done with the help of the rich countries and worldwide organisations.

Countries remain poor when either

◆ they fall into **debt** because the money they earn from their exports is less than the money they have to pay for their imports, or

◆ they borrow money in order to build things like schools, hospitals and roads.

Aid – a traditional method of help

Aid is the giving of resources such as money, medical equipment, food supplies or human helpers to another country. If you or your school are asked to give help it is likely to be for an emergency following a natural disaster such as an earthquake, flood or tsunami. This type of aid is **short term** (photo **A**).

A different type of aid is when poor countries do not have enough money either to buy goods or to improve their services. The only way to get money is to borrow from rich countries or from organisations such as the World Bank. This type of aid, which is **long term**, comes as a loan. As interest has to be paid on the loan, the poor country is likely to fall further into debt. Many of the poorest countries are found in Africa. As they do not earn enough from their exports to pay off their debt they have to go on borrowing. This type of aid widens the gap between the rich and the poor.

Attempts to reduce a poor country's debt

Rich countries are at last realising that many countries will never be able to escape poverty mainly due to their debts – but there are no easy solutions. Five suggestions are given in drawing **B**.

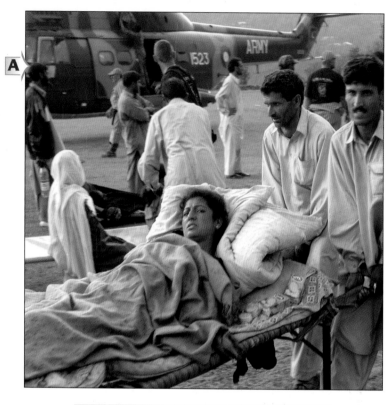

A

How can a poor country's debt be reduced?

Poor countries should be paid more for their exports to increase their income.

Rich countries should reduce the price of their exports to the poorer countries.

B

Rich countries could reduce interest payments on the loans they give to poor countries.

The poorest countries should have their debt cancelled.

Make sure aid goes to the people who need it – not to the Government officials.

Self-help – an alternative approach

Practical Action, which used to be known as Intermediate Technology Development Group, is a British charity. It has a different approach to overcoming poverty in poor countries. It believes that aid is of limited value as it is often only given for a short period of time. Practical Action suggest that it is better to encourage people to look after themselves in order to gain self-respect and independence. It believes in the proverb: 'Give a man a fish and you can feed him for one day (*aid*), teach him how to fish and he can feed himself for life (*education*)'. If he is also taught how to make fishing nets then he becomes independent.

Practical Action has helped people in countries like Kenya through projects such as those shown in diagram **C**. Marietta, whom we read about on page 136, has also been helped by Practical Action (diagram **D**).

C Developing local forms of sustainable energy

PRACTICAL ACTION
Technology challenging poverty

Improving building materials – roof tiles

Improving building materials – mud replaced by a mix of soil and cement

Improving cooking stoves – use less firewood

Marietta was chosen by her community to become a vet. She then received basic training by Practical Action. A five-day course helped her recognise and treat simple livestock diseases and enabled her to give basic animal health care. She was then given an animal first-aid kit.

Each week she spends two mornings in her local 'surgery' and other days visiting local shambas. To reach the surgery Marietta has a four-hour return trek along a track where lions are known to have killed people! Some shambas are even further away.

She earns a small commission from the sale of vaccines and medicine, but does not receive a salary. However, she can now keep her two cows healthy and has some money to buy food in times of shortage.

D

Activities

1 a What is meant by the term 'aid'?
 b What is the difference between short-term aid and long-term aid?
 c Give two advantages and two disadvantages of each type of aid.

2 a Make a copy of drawing **E**.
 b Add labels to show how a charity organisation like Practical Action can help reduce the effects of poverty.
 c Describe how Practical Action has helped Marietta. Use the following headings:
 • Training/education
 • Benefits to herself
 • Benefits to others

E

Summary

There are no easy solutions when trying to reduce poverty in the world's poorest countries. One way is for people to help themselves – this is called self-help.

World issues enquiry

Trade is important in the world because it helps countries share resources and earn money. Unfortunately, not all countries get a fair deal from world trade. For example, poorer countries tend to **export** mainly primary goods which earn little money but **import** mainly manufactured goods that cost a lot of money. This unfair trade is a main reason why poor countries remain poor and many of their people live in poverty.

In this enquiry you work for an organisation that supports **fair trade**. Fair trade can help people in poorer countries to make more money and escape from poverty. This can help improve their standard of living and quality of life. The main aims of the organisation are shown in drawing **B**.

Your organisation has been approached by Kenya Coffee, a company that has decided to support fair trade. Kenya Coffee buys coffee beans mainly from Kenya (photo **A**). It ships the beans to the UK where they are processed and sold throughout Europe. Your task is to explain the main points of fair trade to the company and suggest how it can become a fair trade company.

A

B

Fair Trade

Works for a better deal for Third World producers

- Encourages fair trade between countries
- Aims to reduce poverty mainly by paying higher prices to producers
- Tries to ensure a safe and healthy working environment for workers
- Supports sustainable farming methods and encourages a concern for the environment

How can fair trade help reduce poverty?

1 Look at drawing **C**.

  a How much money goes to the coffee growers?

  b How much money ends up in Kenya?

  c How much money ends up in the UK?

  d If the price of coffee in the shops goes up, where does most of the extra money go to?

2 Look at drawing **D**.
List the points which may help:

  a reduce poverty

  b improve conditions for growers

  c protect the environment

  d affect people in the UK.

Not all of the points in the drawing need be used.

3 Write a report for Kenya Coffee about fair trade, using your answers to activities 1 and 2. Use the headings shown below. Link any suggestions you make to the aims of FairTrade in drawing **B**.

The Fair Trade Report

- The need for **fair trade**
- What a **fair trade** company needs to do
- What the coffee growers will give in return
- The effects of **fair trade** on people living in the UK
- The effects of introducing **fair trade**

- **Growers** work very hard, usually on small farms.
- **Traders** buy from the growers and sell to companies in rich countries.
- **Exporters** organise the transfer of coffee beans from Kenya to the UK.
- **Shippers** transport the beans by sea from Mombasa to London.
- **Processors** roast the beans and prepare them for selling.
- **Retailers** sell the coffee in shops and supermarkets in the UK.

C

Growers	7p
Traders (Kenya)	10p
Exporters (Kenya)	8p
Shippers (UK)	5p
Processors (UK)	38p
Retailers (UK)	32p

If this jar of coffee cost £1 (or 100p) in the shops, this is roughly where the money would go.

D

Pay higher than the world market price for coffee beans.

Pay money to farmers in advance to prevent debt.

We will feel good about helping reduce poverty.

If we are giving the coffee growers all this, in return they must make some promises.

Insist on minimum health and safety standards.

Guarantee equal rights for women and no child labour.

Use sustainable farming methods.

Some jobs may be lost.

Increase the price of coffee in the shops.

Set up processing factories in Kenya.

Ensure decent wages and good housing.

Coffee will cost a bit more.

Minimise damage to the environment.

Buy direct from the growers.

Fair Trade — Works for a better deal for Third World producers

Glossary and Index

Positive factors Things that encourage people to live in an area. *90–92, 104*

Power Energy needed to work machines and to produce electricity. *48–49, 51, 55, 73, 78–79*

Primary activities Collecting and using natural resources, e.g. farming, fishing, forestry and mining. *46–47, 74, 120–121*

Pull factors Things that attract people to live in an area. *97*

Push factors Things that make people want to leave an area. *97*

Quality of life A measure of how happy and content people are with their lives and the environment in which they live. *82, 122, 136, 138, 140*

Quarry A place where rock for building is obtained from the surface of the land. *52, 74–75, 118–119*

Raw materials Natural resources which are used to make things. *46, 48–52, 54, 74, 91, 120*

Recycling Turning waste into something which is usable again. *82–83, 115*

Refugees People who have been forced to move away from their home country and are often left homeless. *99*

Renewable resources Resources which can be used over and over again, e.g. wind. *76–77, 79, 131–132*

Resources Things which can be useful to people. They may be natural like coal and iron ore, or of other value like money and skilled workers. *64–65, 72–83, 130, 137*

'Rich' North Developed countries with considerable wealth and a high standard of living, mainly located in North America, Western Europe, Australasia and Japan. *121–123, 134–135, 138*

Rift valley A deep, steep-sided valley formed by the sinking of land between two faults or cracks in the earth's surface. *110*

River cliff The steep slope cut into the valley side by erosion on the outside of a river bend. *14*

Rural-to-urban migration The movement of people from the countryside to the towns and cities. *97, 113, 132*

Science park An estate of high-tech industries having links with a university. *56–57*

Sea defences Features added to a coast to protect it from erosion and flooding. *20–21, 129*

Secondary activities Industries that make or manufacture things by processing raw materials or assembling parts to make a finished product, e.g. steelmaking and car assembly. *46–47, 120–121*

Service industries Activities that aim to help people, e.g. teaching, nursing. They are also known as tertiary industries. *46–47, 120–121*

Set-aside land A scheme where farmers are paid for not farming their land. *35, 38*

Sewage Waste material from homes and industry. *114–115, 132, 137*

Shanty settlement A collection of shacks and poor-quality housing which often lack electricity, a water supply and sewage disposal. *114–115, 118*

Short-term aid A type of aid usually given by richer countries and voluntary organisations to poorer countries immediately after a natural disaster such as an earthquake or a tsunami. *138*

Silt Fine soil left behind after a river floods. Also called 'alluvium'. *15, 24*

Site The place where a settlement or a factory is located. *48–58, 60–63*

Soil erosion The removal of soil by wind or water. *37*

Sparsely populated An area that has few people living in it. *88–92, 112*

Spit A long narrow accumulation of sand and shingle that grows out from the coastline. *17*

Stack A detached pillar of rock on a sea coast separated from the mainland by erosion. *16*

Standard of living How well-off a person or a country is. *108, 120, 122, 136, 138, 140*

Subsidies Money paid to farmers by the government or the European Union. They usually come in the form of grants and loans. *31, 33, 35*

Sustainable development A way of improving people's standard of living and quality of life without wasting resources or harming the environment. *66, 82–84, 118–119, 139*

Sustainable energy A way of providing affordable energy that will improve people's quality of life without spoiling the environment in which they live. *77, 83, 139*

Tertiary industries Activities that provide a service for people, e.g. teaching, nursing, retailing. *46–47*

Thermal power Electricity produced by burning fossil fuels. *78–79*

Trade The movement and sale of goods between countries. *70, 120–123, 138, 140–141*

Transport Ways of moving people and goods from one place to another. *41, 47–52, 55, 58–61, 127, 137*

Transportation The movement of material by rivers, sea, ice and wind. *9–10, 17*

V-shaped valley A valley which has been eroded by a river so that its cross-section looks like the letter V. *10–11*

Waterfall A sudden fall of water over a steep drop. *12–13, 22–25*

Water supply The availability of a continuous supply of clean water. *91, 112–115, 132–133, 137, back cover*

Waves Formed when wind blows over the sea. *8, 16, 18–21, 77*

Weathering The breakdown of rocks by climate, chemicals, plants and animals. They are broken down without being removed. *6–8, 10, 19*

Wetlands Marshy areas which are a habitat for wildlife. *36, 68–69*

Wildlife habitats The homes of plant and animals. *35–36, 68-69, 80-81, 84–85, 129*